The Ocean World of Jacques Cousteau

Volume 1
Oasis in Space

THE DANBURY PRESS

The Danbury Press
A Division of Grolier Enterprises Inc.

Publisher: Robert B. Clarke

Published by The World Publishing Company

Published simultaneously in Canada
by Nelson, Foster & Scott Ltd.

ISBN 0-529-04936-8
Library of Congress catalog card number: 72-87710

Printed in the United States of America

456789987654

Project Director: Peter V. Ritner

Managing Editor: Steven Schepp
Assistant Managing Editor: Ruth Dugan
Senior Editors: Donald Dreves
Richard Vahan
Assistant Editor: Sherry Knox

Creative Director and Designer: Milton Charles

Assistant to Creative Director: Gail Ash
Illustrations Editor: Howard Koslow

Production Manager: Bernard Kass

Science Consultant: Richard C. Murphy

"This little lionfish nuzzling the reef for food has been caught and classified many times by marine biologists in the past 200 years. We know more about his anatomy than about his habits. We do know that he carries venom in some of his dorsal spines that could be extremely toxic to man, but that he only uses this for emergency defense. What does he sense with his little brain, what does he feel within his little body? Is all his life motivated only by simple impulses, like hunger or sexual drive? Is he more than a beautiful mechanical toy? I know there will be no answer to such questions in my lifetime. But the mystery of life is as inspiring as music or poetry."

Contents

In the oceans have been tested all conceivable BEHAVIOR PATTERNS (Chapter VII). Some of these patterns are protective, some destructive: a dolphin could very well survive alone, but if isolated from his companions he prefers to die. Nature displays every combination.

Most of the time the oceans are in a state of "armed truce" interrupted by brief periods of hunting and violence. The complexity of the relationships has created an intricate, fragile, but stable *balance:* INTERDEPENDENCE (Chapter VIII).

At the turn of the century, when Prince Albert I of Monaco brought back to his oceanographic vessel the first catch of hundreds of abyssal fish from 20,000 feet of depth; when Jacques Piccard and Don Walsh in their historic dive in the bathyscaph *Trieste* witnessed life at the bottom of the Mariana Trench; when Albert Falco steered diving saucers on the continental shelf—they were just beginning to understand the SEASONS AND PROVINCES OF THE OCEAN (Chapter IX).

But in the sunny layers of the tropical sea, innumerable small polyps have built structures that dwarf the most ambitious man-made constructions: THE PHARAOHS OF THE SEA (Chapter X).

Man is turning today to the RICHES FROM THE SEA (Chapter XI). But some of these resources have limitations: overfishing today depletes reserves: offshore oil is not inexhaustible; whaling brings about extinction of the largest animal ever to have existed.

Man's RETURN TO THE SEA (Chapter XII) is dramatically illustrated by the historical telephone conversation in 1965 between Philippe Cousteau and Scott Carpenter.

Because of overfishing, mismanagement, ruthless destruction, and pollution, today THE SEA IS IN DANGER (Chapter XIII). Obviously, an individual man is intelligent enough to understand that he is not a dinosaur, but *as a species* he may not be responsible enough to avoid the fate of the dinosaur.

Frightened by his concrete jungles, man turns to the oceans for leisure and escape, only to find that they are dying from his carelessness. Man takes off for the conquest of space, only to find that the solar system is a dustbin of dead celestial bodies. The truth is that man is alone—a lonely, pulsating, thinking creature—in the universe. Alone on board his Spaceship Earth, an OASIS IN SPACE (Chapter XIV), his only cousin is the dolphin, the only creature with which he can hope one day to communicate.

Introduction: If the Oceans Should Die

It is a shocking paradox that at the precise moment in history when man is arriving at an understanding of the sea he should also have to face the question above. Just now, in our generation, when after many thousands of years of ignorance and superstitions man is at last beginning to learn about managing and exploiting the vast resources of 70 per cent of earth's surface, he finds himself in a race against time to rescue it from his own spoilations.

If the oceans of earth should die—that is, if life in the oceans were suddenly, somehow to come to an end—it would be the final as well as the greatest catastrophe in the troublous story of man and the other animals and plants with whom man shares this planet.

To begin with, bereft of life the ocean would at once foul. Such a colossal stench born of decaying organic matter would rise from the ocean wasteland that it would of itself suffice to drive man back from all coastal regions. Far harsher consequences would soon follow. The ocean is earth's principal buffer, keeping balances intact between the different salts and gases of which our lives are composed and on which they depend. With no life in the seas the carbon dioxide content of the atmosphere would set forth on an inexorable climb. When this CO_2 level passed a certain point the "greenhouse effect" would come into operation: heat radiating outwards from earth to space would be trapped beneath the stratosphere, shooting up sea-level temperatures. At both North and South Poles the icecaps would melt. The oceans would rise perhaps 100 feet in a small number of years. All earth's major cities would be inundated. To avoid drowning one third of the world's population would be compelled to flee to hills and mountains, hills and mountains unready to receive these people, unable to produce enough food for them. Among many other consequences of the death of the oceans, the surface would become coated with a thick film of dead organic matter, affecting the evaporation process, reducing rain, and starting global drought and famine.

Even now the disaster is only entering its terminal phase. Packed together on various highlands, starving, subject to bizarre storms and diseases, with families and societies totally disrupted, what is left of mankind begins to suffer from anoxia—lack of oxygen—caused by the extinction of plankton algae and the reduction of land vegetation. Pinned in the narrow belt between dead seas and sterile mountain-slopes man coughs out his last moments in unutterable agony. Maybe thirty to fifty years after the ocean has died the last man on earth takes his own last breath. Organic life on the planet is reduced to bacteria and a few scavenger insects.

Why begin a work on the subject I love most in the world with this nightmare? Because the ocean can die—and because we want to make sure that it doesn't. Man exists only because his home-planet, Earth, is the one celestial body we know of where life is at all possible. And life is possible on earth because earth is a "water planet"—water being a compound itself probably as rare in the universe as life, perhaps even synonymous with life. Water is not only rare, not only infinitely precious—it is peculiar, with many oddities in its physical and chemical make-ups. It is out of this unique nature of water, interacting with the dynamics of the world "water system," of which the sun and the ocean are the motors, that life originated. The ocean is life.

This is why we must change our attitudes toward the ocean. We must regard it as no longer a mystery, a menace, something so vast and invulnerable that we need not concern ourselves with it, a dark and sinister abode of secrets and wonders. Nor do we want to follow the methods of the first scientists who sailed and searched the seas to compile lists: lists of mammals, lists of sea-birds, of jelly-fish, of temperatures, of currents, of migratory patterns. Instead we want to explore the themes of the ocean's existence—how it moves and breathes, how it experiences dramas and seasons, how it nourishes its hosts of living things, how it harmonizes the physical and biological rhythms of the whole earth, what hurts it and what feeds it—not least of all, what are its stories.

I dedicate these books to water and the life which depends on water—and to the mother of waters: the Ocean.

Jacques-Yves Cousteau

Chapter I. Water–The Essence of Life

Of all the earth's resources, water is the most precious. Without it no plant or animal could have evolved.

Where did the water come from that makes life possible? The most widely accepted theory proposes that water in the form of vapor was one of the compounds present in very small amounts in the vast gaseous cloud of hydrogen and helium from which our sun evolved. The sun became a separate body without using up all the matter in the cosmic

> "Every drop of water on earth when earth was formed remains on our earth today."

cloud. What remained whirled through space and after millions of years became the nine planets of our solar system. The water from the original cloud, according to this theory, became a part of each planet, but was not necessarily evenly distributed. The amount of water present and the form it took—solid, liquid, or vapor—depended on the planet's mass and its distance from the sun. The moon, for example, is too small to retain an atmosphere or water vapor.

Through the happiest of accidents, the earth is the right size and distance from the sun to permit water to exist in all three forms. It is in its liquid form that water is essential to the life process. And it is in liquid form that water is rare in the universe. Close to the sun the heat is so intense that water is vaporized. Far from the sun it is so cold that water remains permanently frozen. Of the other planets, only Mars, our second nearest neighbor, is in the narrow temperature band in which water can exist in its three states. Thus, extraterrestrial life in our solar system is deemed possible only on earth and

maybe Mars. Whether life, in fact, exists on Mars we do not yet know.

Every drop of water on earth when earth was formed remains on our planet today. It has since gone many times around the globe; water is constantly moving. It is pulled from the oceans by the heat of the sun and moved along in clouds by winds. It may fall from the sky as rain, snow, sleet, or hail. It may fall in any latitude from Arctic to Antarctic. It will stay for ages in icecaps and glaciers, or return promptly to the sea through storm sewers and rivers. Almost all of the water on earth today is in the oceans. Life itself is thought to have begun there. No one is sure exactly how. But at an unknown time, probably 3 billions of years ago, some fortuitious circumstance made it possible for chemical substances dissolved in the warm broth of the sea to combine with each other to form molecules with the unique ability of reproducing themselves. From those first living molecules all life has descended.

A / Ice. Captain Cousteau and his team preparing to film the walrus in the semi-frozen waters of Alaska.

B / Fresh water. Deep-sea divers in a fresh-water grotto.

C / Oceans. Earth as seen from space.

D / Clusters of clouds. Yet another form water takes.

Chapter II. A Man Is Born—A Dolphin Is Born

A man is born. The most elaborate creature on earth is frail, he has no natural weapons for attack or defense. For years he will need the constant care of his mother, the protection of his father. All his life his body will have to fight gravity. He is a visual creature, sight being his most crucial sense. He is gifted with the ability to laugh, to play, to make love in all seasons. He is a social animal, but is often aggressively violent toward his own kind. He has access to the knowledge accumulated by his ancestors thanks to written or electronic storage of information. This baby faces a destiny that can range from the miserable condition of an urban slum dweller to the fascinating life of an outer-space explorer.

A dolphin is born. The most elaborate creature in the sea also needs the care of his mother for about a year. He is weightless in the sea so that most of his muscles can be devoted to propulsion. His life-expectancy is half that of his human counterpart. He is an acoustical creature, capable of "seeing with his ears" much farther than with his eyes. He plays most of his time, has an active sex life all year round. His social rules are severe, but he is not aggressive. He probably communicates by a modulated language, but we don't know yet to what extent he can store and transmit information. This baby's destiny is one of an easy, happy life, sometimes complicated by parasites, rarely endangered by natural enemies.

The Dolphin—A Marine Playboy?

The dolphin's life is a little like that of a human playboy's. He doesn't live as long; 30 is probably the maximum age for most species of dolphin. But he moves into the interesting business of life much more swiftly than the man. He has a gestation period of 12 months and a lactation period of another 9 months—somewhat similar to the human rhythm. As with man the dolphin protects its young from danger. To ward off shark attacks a group of adults may encircle the young as a barrier.

A dolphin is sexually mature at eight years —and the female can give birth once a year after this for another 14 years or so. In their free natural habitat dolphins tend to breed in spring and summer. Early morning seems to be the favorite mating hour. But the animal is sexually active all year round. Sexual experimentation of all kinds, body-rubbing, mouthing, biting, flipper-touching form a big part of the dolphin's life. And he has plenty

> **"The dolphin prefers to play. When all else failed to amuse him, the dolphin invented body-surfing."**

of opportunity for it. For he is an awesomely efficient swimming machine, equipped with acoustic range-finders and a capacious brain, who can count on providing himself a daily diet without wasting much time on it.

With a body beautifully streamlined to minimize resistance while coursing through the oceans, and a powerful tail that makes it possible for him to stand two-thirds of his body's length out of the water for seconds at a time ("sculling"), the dolphin can if required swim at a velocity of 20 miles an hour almost indefinitely. He is also an expert diver. His blood can absorb more oxygen than man's can and he has additional oxygen stored in muscle reservoirs. By slowing his heart beat and restricting the flow of blood except to the brain he makes the best use of this oxygen in his dives. His tissues can tolerate an oxygen debt and his brain is less sensitive to CO_2 than ours. His lungs are undamaged by complete collapse. As a result the dolphin can easily dive deeper than 300 feet and remain underwater for more than five minutes. Some have reached depths of 1000 feet—where the naked human diver rarely dives to 100 feet.

These remarkable adaptations and the dolphin's physical appearance point up the long years that have elapsed since dolphins separated from their terrestrial ancestors and embarked on their zoological adventure in the undersea—to become one of the most specialized and efficient marine creatures that has ever existed. But this all sounds like work. The dolphin prefers to play. When all else failed to amuse him, ages ago, the dolphin invented body-surfing—and in many parts of the world can be seen rollicking in the breakers in groups of four or more. With the advent of the steamship the dolphin discovered another game — riding the bow waves. Perfectly calculating the speed of the ship, the contours of waves, winds, and currents, one or more dolphins take up a position in front of a fast-moving boat and, with next to no expenditure of energy on their part, hitch a ride. Let man do the work!

Unlike us, the dolphin has found no need to develop his territory by constructing huge habitation complexes. He is content to play, perpetrate his species, and take from nature only what he needs. Such an animal has no enemy in the sea except for his cousin the killer whale. And, alas, man, who cheats— with net, harpoon, and gun!

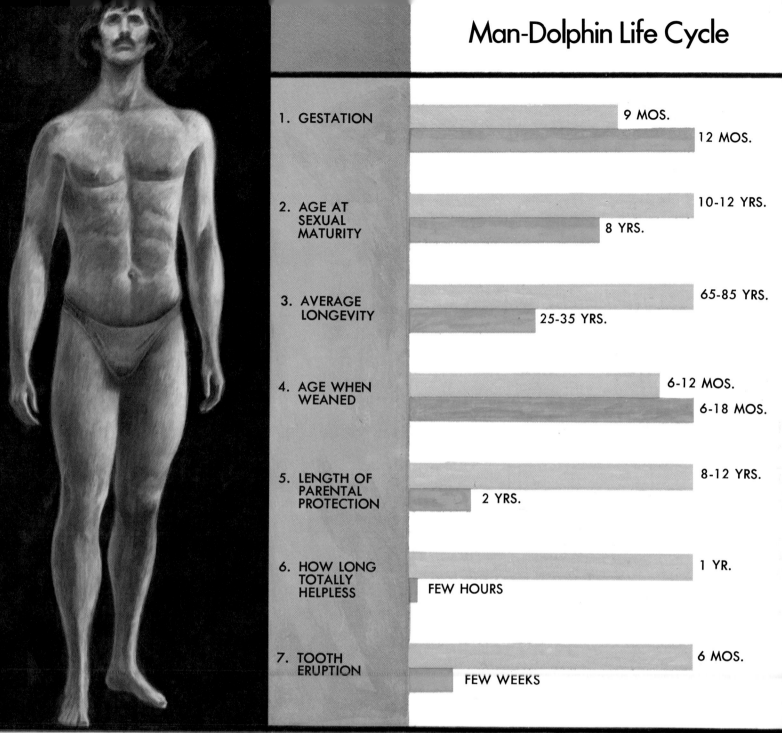

Man-Dolphin Life Cycle

	Man	Dolphin
1. GESTATION	9 MOS.	12 MOS.
2. AGE AT SEXUAL MATURITY	10-12 YRS.	8 YRS.
3. AVERAGE LONGEVITY	65-85 YRS.	25-35 YRS.
4. AGE WHEN WEANED	6-12 MOS.	6-18 MOS.
5. LENGTH OF PARENTAL PROTECTION	8-12 YRS.	2 YRS.
6. HOW LONG TOTALLY HELPLESS	1 YR.	FEW HOURS
7. TOOTH ERUPTION	6 MOS.	FEW WEEKS

Chapter III. Cradle of Life

Ours is a watery planet. The oceans, which cover almost three quarters of the earth's surface, contain some 330 million cubic miles of water. Dissolved in that huge liquid expanse are a variety of salts and minerals

"A 'dry water cycle' took place millions of times before the earth cooled enough to permit the water to remain and accumulate in cracks and depressions."

and the gases necessary to the teeming and almost unimaginably varied life in the sea. Dissolved oxygen is breathed by aquatic animals. Dissolved carbon dioxide is extracted by marine plants and used by them to produce food.

There was no sea during the earth's infancy while the planet's crust was forming and cooling. At an early stage water released from our hot planet vaporized and rose to envelop it in a vast cloud. This water cooled, condensed and fell to earth during almost permanent thunderstorms, only to be revaporized and rise again. This "dry water cycle" took place millions of times before the earth cooled enough to permit the water to remain and accumulate in cracks and depressions.

If we trace the long infinitely slow process of evolution backwards, we would begin with the two highest creatures yet to evolve: man, the most intelligent creature on land; and the dolphin, the most intelligent creature in the sea. Both man and dolphin belong to the same animal grouping—both are warm-blooded, air-breathing mammals, vertebrates that nurse their young. Man and dolphin are both descendants of the early

mammals that roamed the land—sea creatures who for a time lived on the continents and then returned to the sea. These prehistoric mammals were heirs of cold-blooded vertebrates who had evolved from the first fish that managed to crawl from the water and adjust to life on land. This transition involved adjusting to the full force of gravity, not felt underwater, and obtaining oxygen from air instead of water. Those air-breathing and walking fish in turn had evolved, through many steps and millions of years, from primitive animals without backbones who were distant descendants of one-celled animals. But even before the animal world of the oceans could evolve animal life and plant life had become differentiated. Organisms that acquired the capacity to use energy from the sun to manufacture food for themselves out of the chemicals dissolved in the sea became specialized as plants.

"It is one of man's illusions that the oceans and land masses of the world are eternal."

It is one of man's illusions that the oceans and land masses of the world are eternal and that he himself is the ultimate end product of billions of years of evolution. Though there are reasons to believe that the rate at which new forms of life are produced has substantially slowed down, the earth remains in a constant state of flux. In the process of evolution which continues from moment to moment, the human species interferes dramatically. Today, his influence is tragically destructive. But we can hope that Science and Reason will reverse the trend. In any case, the world of the future will surely be as different from the world of today as today's world is from that of the past.

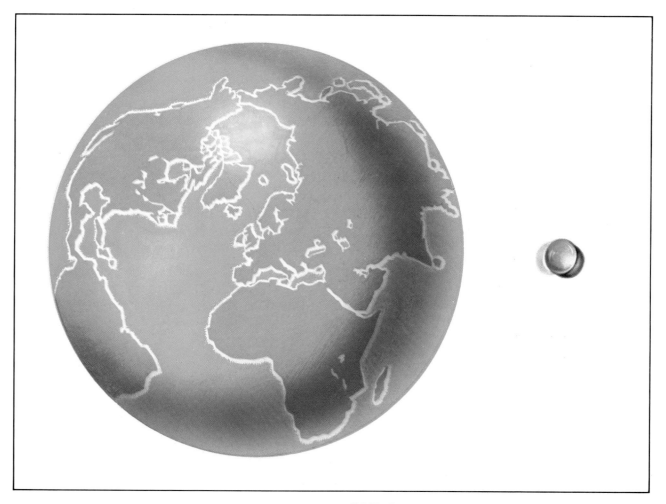

*"**All the water** on earth makes up only a little more than one-tenth of 1 per cent of the earth itself."*

Solid Earth v. Liquid Ocean

Were the crust of earth to be levelled—with great mountain ranges like the Himalayas and ocean abysses like the Mariana Trench evened out — no land at all would show above the surface of the sea. Earth would be covered by a uniform sheet of water—more than 10,000 feet deep! So overwhelming the ocean seems to be. The earth is truly a water planet. Yet there is another way of looking at it, perhaps more appropriate when thinking of man. All the water on earth, fresh and salt, makes up only a little more than one tenth of 1 per cent of the volume of earth itself.

In this perspective one understands why we must conserve and cherish this limited and vital resource. Water makes our earth livable. It is, in fact, life itself. Water modifies our climate by storing heat and redistributing it over the earth through ocean currents and atmospheric circulation. Water is the universal solvent — allowing more substances to be dissolved in it than any other compound. In man water is the vehicle which transports vital materials throughout the body, making possible life's complex biochemical reactions. Water is indeed a rare and precious gift!

17

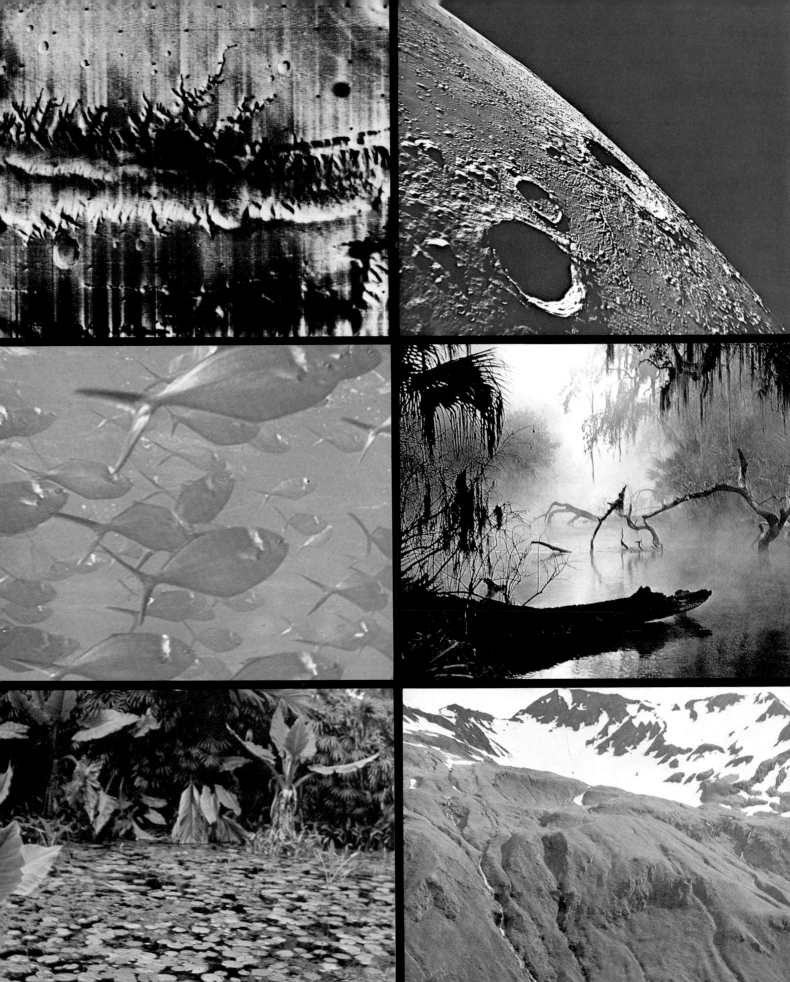

Water Is Precious

Water, along with sunlight and oxygen, is one of the prime ingredients that make life on earth possible. Without water life as we know it just can't be. So water is one of our most precious assets. Deserts are virtually waterless and virtually lifeless too. Oceans, on the other hand, are virtually all water, and the oceans teem with life.

A / Mars. *The canyons and terraces suggest the presence of water—is there life there too?*

B / The sea. *A lively school of jacks.*

C / Fresh-water marsh. *Abounding with plant and animal life.*

D / The Moon. *Without water—dead dust and rock.*

E / Mangrove swamp. *Where marine life begins in the tropics.*

F / Mountains. *Water in both its solid and liquid states—depending on altitude.*

G / Desert. *Waterless, and thus nearly lifeless.*

H / Grasslands. *An abundance of water—an abundance of greenery.*

I / Polar icecap. *Life near and about water even in extremes of temperature.*

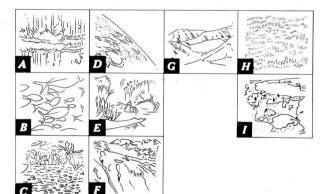

Calypso at Dawn

"It is early morning. I am alone on the deck of the research ship *Calypso*. The sea is flat and grayish-blue. The air is hot and sultry. There is hardly any breeze. Overhead, heavy dark clouds are silently piling up into billowing pyramids. We are moving through a belt of water slightly north of the equator known as the doldrums. In the days of wind-driven sailing ships mariners feared these waters. The doldrums is the region where the trade winds never blow. It is where ships used to becalm and where supplies would run out. But I hear the steady throbbing of *Calypso's* engines. I know this ship will pass through the doldrums without mishap.

"What is it about the sea, I wonder as I stare into its depths, that has always lured man on to explore its secrets? I think of the great voyagers of the past—Ericson, Magellan, Vasco da Gama, Francis Drake. Their daring navigation told us much about the size and shape of the oceans of the world. Today we know the configuration of the surface of the sea with an accuracy they could not have dreamed of.

"It is the underwater world we now explore. We are looking there for clues to our own distant past, for information about the formation of our planet and the origin of life. *Calypso* is crammed with equipment to aid us in that search. Our underwater observation chamber enables us to watch and photograph marine creatures wherever we venture. Divers going down to depths of 200 feet for a firsthand look at the ocean's habitats enter and leave the water through our midships' diving-well. Our echo-sounders help us map the ocean floor. Other research vessels have drills that bore into the earth's crust and bring up cylindrical samples of that ancient material. We are like detectives trying to unravel what happened millions and billions of years ago.

"Somewhere on the ocean floor are bits and pieces telling of the story of this planet's beginning. It is in the oceans, I believe, rather than the skies that we will finally learn how earth and life formed. And the sea, which gave birth to us all, will shelter us once again. Looking down at the water I can imagine man-made islands of the future floating offshore and other settlements rising from the ocean floor. Before man leaves earth for an-

> "Where did this bountiful planet come from? Are there other planets anything like ours in other solar systems? Do they too, if they exist, have comparable forms of life?"

other home I am sure he will retrace his steps through the sea.

"Raindrops begin to fall. They too come from the ocean, I remember, as I climb the steps to *Calypso's* bridge. It is at peaceful moments like this that I ask myself the questions I am sure man has been asking himself without answers for almost as long as he has inhabited the earth. When you raise your eyes to the sky, you inquire—where did this bountiful planet come from? Are there other planets anything like ours in other solar systems? Do they too, if they exist, have comparable forms of life? Or life we cannot imagine? Where did the spark of life on this planet come from?"

Calypso—originally a minesweeper built in America which served in the Royal Navy and after World War II was converted to one of the great ladies of ocean science: tough and tireless.

▲ HUMPBACK WHALE

Chemical Content of Human Blood and Ocean

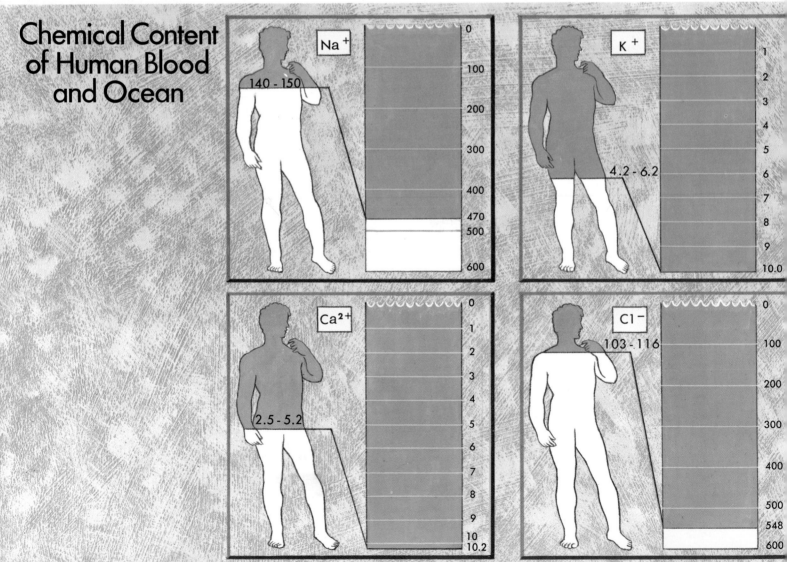

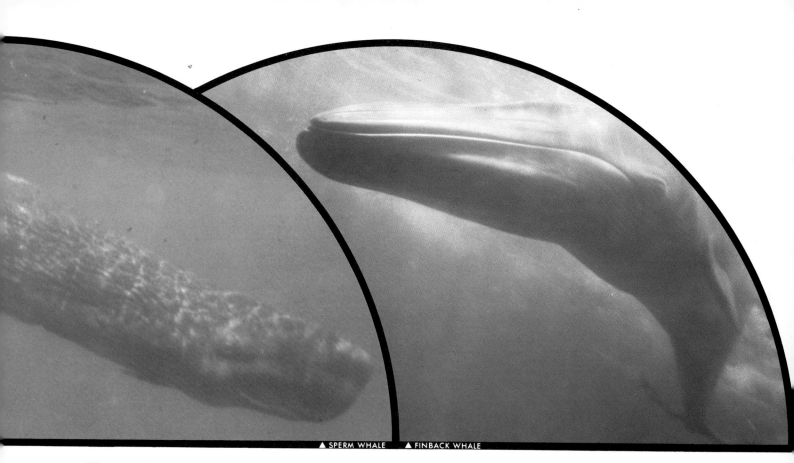

From Land to Sea

Whales possess reduced bones representing four limbs, hips, and an articulated neck. Unlike fish, they lack bones in the caudal fluke and dorsal fin. The story of whales most probably began with a primitive land-dwelling mammal 125 million years ago. This creature of the distant past may have found it advantageous to search for food in marshes and estuaries rather than compete on land for mice or lizards. At one stage our primitive whale may have looked similar to a seal. Millions of years of trial and error culminated in the contemporary cetacean, highly adapted to the marine environment but helpless on land. In addition to the whale's other physical features he now possesses an underwater communication system.

In the diagram to the left, chemical symbols are used. "Na" stands for sodium, "K" stands for potassium, "Ca" for calcium, and "Cl" for chlorine. The four pictures show the relative amounts of these elements in sea water and in human blood.

From Sea to Blood

Today we are pretty sure that life originated in the sea. But around the turn of the century the provocative suggestion was made that man might be carrying a proof of this around in his own circulatory system. It was noticed that there is a similarity in the salt contents of human blood and sea water—especially in the electrolytes most crucial to life: sodium, potassium, calcium, and chlorine. The salts in sea water are at about three times the concentration you find them in blood, but the ratios between these ions are intriguingly alike. We know that sea water has been growing saltier through the eons. Could human blood be an echo of the primitive ocean that was less salty, the ocean as it existed when vertebrates first appeared and established their circulatory systems? The theory has recently come in for hard knocks because it has been demonstrated that even the primitive ocean at the time vertebrates emerged was twice as salty as blood.

Half-and-Half Existence— Land and Sea

Walruses live around the world—but only in the Arctic. Although there's only one species recognized by scientists, they divide them into an Atlantic and a Pacific race. The Atlantic race lives along the seacoast, often on icefloes, in the Canadian Arctic, Greenland, and in European Arctic waters east to Siberia. The Pacific walrus comes from Alaskan and Siberian Arctic coastal waters. Walruses dig clams out of the ocean floor down to a depth of 240 feet with tusks which are overgrown canine teeth that may reach a length of 20 inches. Then they apparently suck the clams out of their shells.

Male walruses get to be 12 or 14 feet long and may weigh up to 3300 pounds. Females get to 10 or 12 feet in length and a ton in weight. They give birth to single pups every two or three years in the spring. The pups weigh over 100 pounds at birth. Walruses live for about 30 years. Their biggest enemy is man, mostly the Eskimo who lives off the walrus's three-inch layer of blubber, flesh, skin, bone, and tusks. But they are protected· from overhunting now. A few years ago a census revealed about 25,000 Pacific walruses and about 30,000 Atlantic walruses left in the world.

Between Two Worlds

"In between two worlds — ours and the abyssal depths — a solitary nautilus has come up to the 200-foot level to feed. I can scarcely believe it! Here is the elusive survivor of hundreds of millions of years beneath the sea! Studied mostly in shell and fossil form, and briefly in captivity, here at last in the wild is the *living* animal. A living fossil! Shy of light, he slowly moves away towards his secret home in the depths. Jet-propelled, he swings rhythmically like the pendulum of an eternal clock. The nautilus has ticked off many ages of evolution. It has seen men returning to the sea as it saw the dolphin return to the sea."

◄ *The common jack or crevalle is a fast-swimming, tireless hunter.*

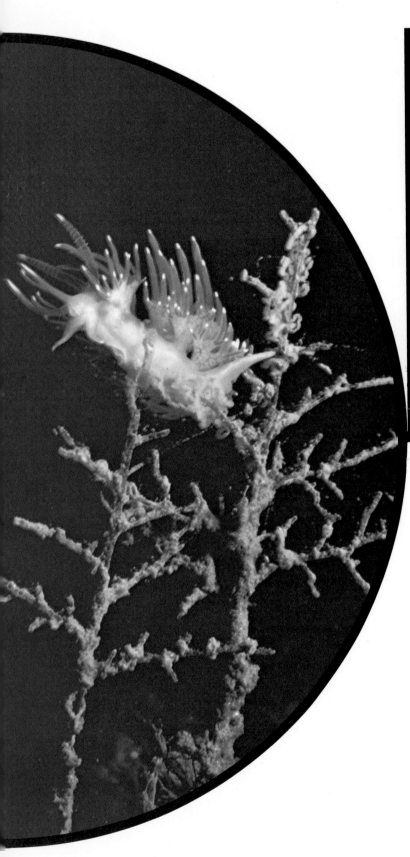

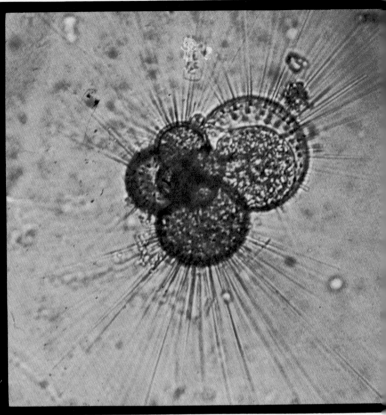

Beginnings Of the Web

Phytoplankton (plants) and zooplankton (animals) form the base of the ocean's food web. Raw materials otherwise useless to zooplankton—or any other animal—are assembled by plants into usable food. This planktonic conversion of material requires sunlight and takes place only in a well-lit, thin, upper layer of the sea. After the phytoplankton have utilized the sunlight, water, and the combination of chemicals available to them, they serve as food for the tiny drifting animals of the sea. These in turn are food for still larger organisms who are food for still larger animals. The many food chains strung together in this manner are linked to each other to form an enormously complex food web.

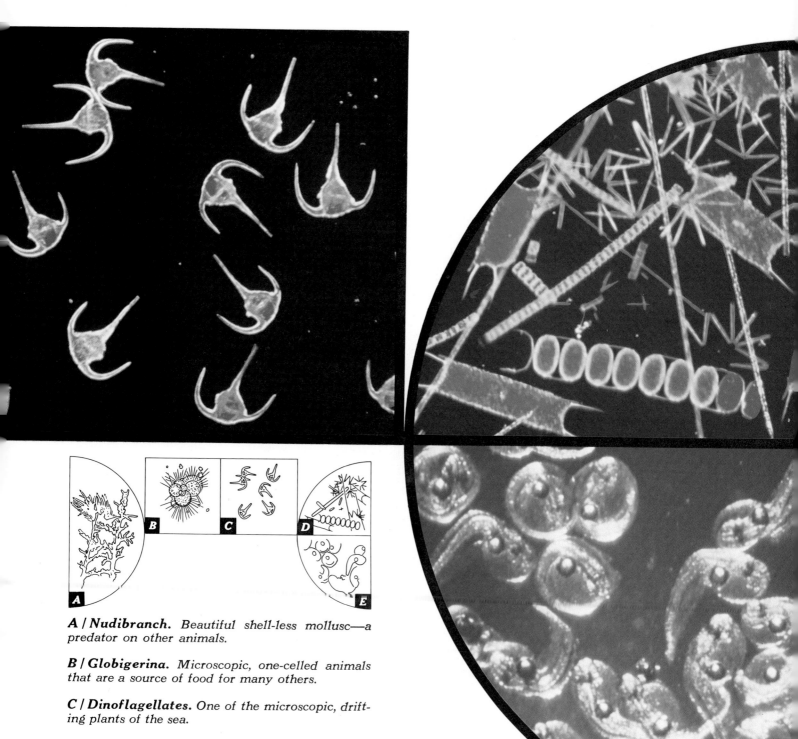

A / Nudibranch. Beautiful shell-less mollusc—a predator on other animals.

B / Globigerina. Microscopic, one-celled animals that are a source of food for many others.

C / Dinoflagellates. One of the microscopic, drifting plants of the sea.

D / Diatoms. Probably the most important basic nutrient provider in the sea.

E / Fish eggs. Helplessly drifting animal matter serving as food for larger organisms.

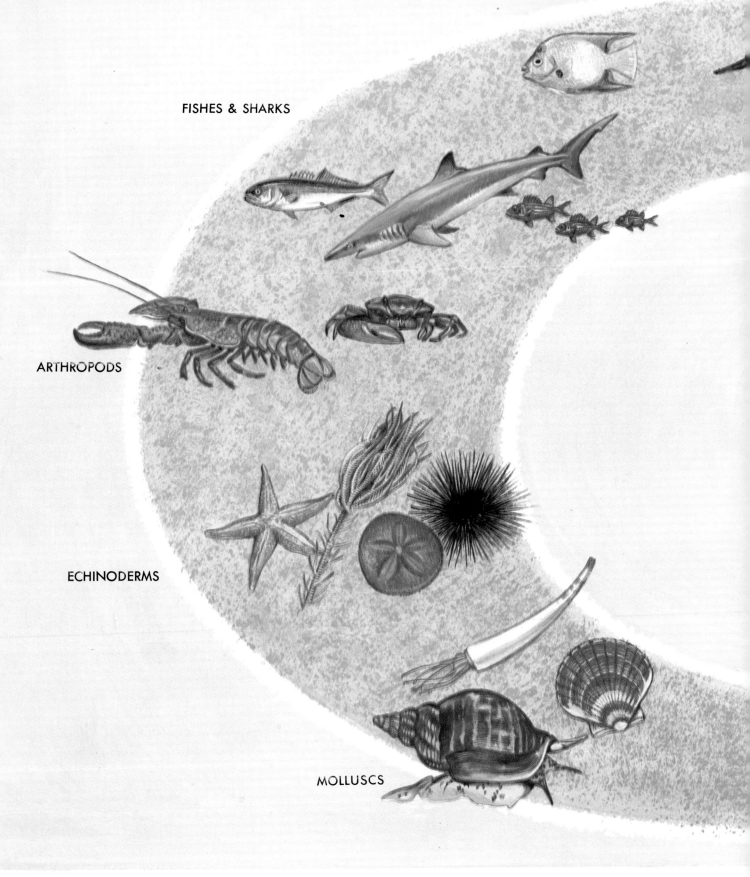

FISHES & SHARKS

ARTHROPODS

ECHINODERMS

MOLLUSCS

The Animal World

All life in the Animal World is related. All life, the most primitive as well as the most

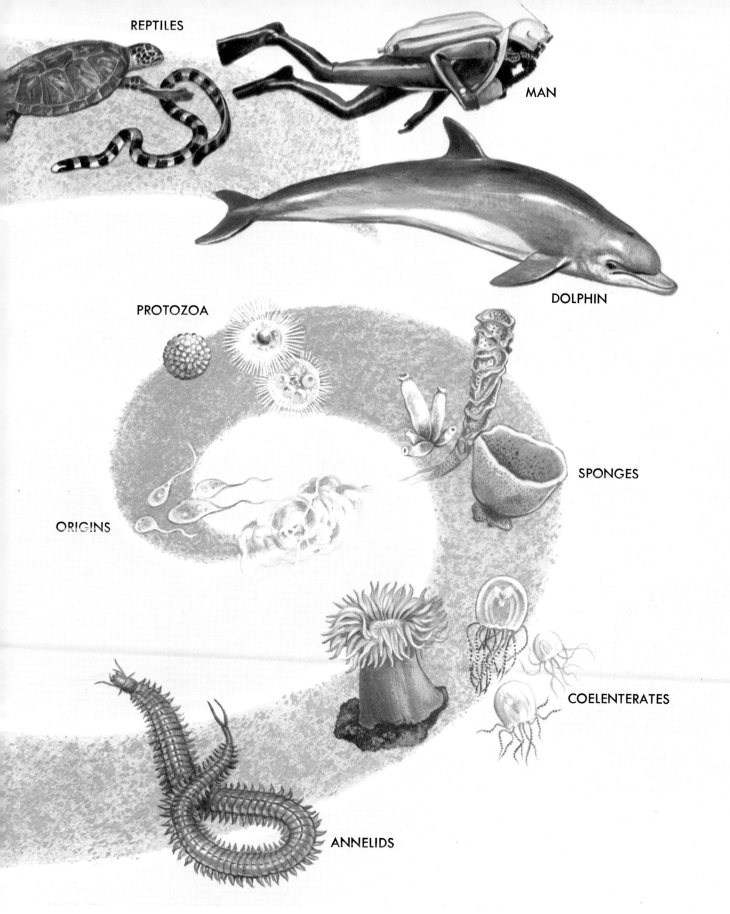

REPTILES

MAN

DOLPHIN

PROTOZOA

ORIGINS

SPONGES

ANNELIDS

COELENTERATES

complex, stems from a common origin. All living things depend upon the same heredi- tary material, the common thread that binds animals and plants together.

31

Chapter IV. Essential Drives

The fundamental law of the animal world in the sea, as on land, is: survive. To that end animals must eat and must procreate. Only if it gets enough food can an animal grow to maturity. And only when mature can animals reproduce and ensure survival. Sea animals feed on each other in a set order

> "Half of 1 per cent of the solar energy falling on the ocean remains in it due to ocean plants."

that moves upward from the lowest to the highest. For all but the swiftest and most powerful, the general rule is eat and be eaten. But with great ingenuity and ferocious energy and tenacity, animals at all levels instinctively protect themselves so the act of reproduction can take place.

Man, though an animal himself, subject to hunger, thirst, the need for shelter, the urge for sexual gratification, and the will to live, has learned through centuries of civilization to tame his instincts and defer his pleasures. Many of man's basic needs are today readily fulfilled. It is thus not easy for modern man to imagine the blind power of the basic drives that move the animal world through its cycles. Other animals act instinctively. It is only in extreme circumstances, when his very life is at stake, that modern man can appreciate the force of the drive to survive.

Survival for the individual is, of course, finally impossible. Animals and men must die. And nature is unsentimental about when and how. But for all the indifference and brutality of life in the wild, it contains little unnecessary slaughter. Animals kill to eat. Dolphins, which like man, have leisure time, spend it peacefully. Man, however, the most intelligent of creatures, is now a threat to himself and to all other life on the earth.

The food chain which keeps the Animal World going is energized by the sun. Half of 1 per cent of the solar energy falling on the ocean remains in it due to ocean plants, mainly the unicellular algae in the plankton, which are the "primary producers." Tiny herbivores eat these plants, tiny primary carnivores eat the herbivores, larger secondary carnivores may eat both primary carnivores and herbivores. And so it goes—with all life of the sea playing its part: the sperm whale chases giant squids through dark waters nearly a mile down; the mako shark slashes at a school of mackerel; the northern fur seal waits for the evening rise of lantern fish; almost imperceptible organic particles slowly drift down to the glass sponges and brittle stars of the abyss. But if the simpler forms spend almost all their lives searching for as much food as they can get, on the contrary, the highly evolved creatures can sat-

> "If simpler forms spend almost all their lives searching for as much food as they can get, highly evolved creatures can devote time to more complicated occupations-play and social relationships, affection and love."

isfy their needs easily and thus devote time to more complicated occupations—play and social relationships, even affection or love.

The Food Chain. Paradoxes abound in the food chain in the sea as on land. Like the elephant on the dry land, the mightiest animals of the sea—the blue whale among the mammals and the whale shark among the fishes—are not the bloodthirsty predators their size and strength would seem to entitle them to be. Huge as they are, they are gentle browsers among the smallest creatures of the sea, the minuscule inhabitants of the plankton layers.

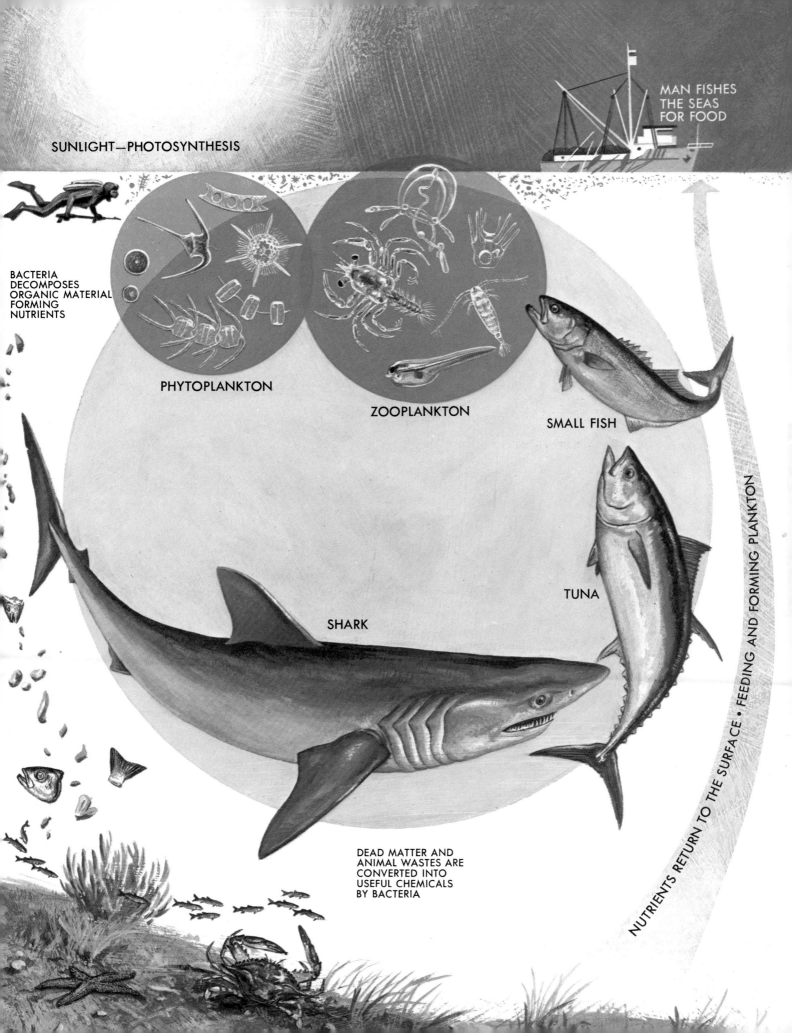

SUNLIGHT—PHOTOSYNTHESIS

MAN FISHES THE SEAS FOR FOOD

BACTERIA DECOMPOSES ORGANIC MATERIAL FORMING NUTRIENTS

PHYTOPLANKTON

ZOOPLANKTON

SMALL FISH

TUNA

SHARK

DEAD MATTER AND ANIMAL WASTES ARE CONVERTED INTO USEFUL CHEMICALS BY BACTERIA

NUTRIENTS RETURN TO THE SURFACE • FEEDING AND FORMING PLANKTON

Catching a Meal

A / Feather duster worm. The feather duster worm has no means of locomotion and must patiently wait for food to drift by.

B / Gorgonian. The polyps of these gorgonian corals possess delicately branched tentacles which provide an increased surface for snaring drifting prey.

C / Anemone. Although the tube anemone cannot change its location the slender tentacles are very active—waving in a gentle current to capture a meal.

D / Starfish. The starfish is an active feeder (in a slow sort of way) which seeks out mussels. Tiny suction cups hold the mussel until it tires and opens up.

E / Grunt. As with many fish, grunts generally feed at dawn and dusk. For them night is not a time of feeding. And during the day food is not available.

F / Grouper. Unlike some other predators the grouper only has small teeth; he uses instead grinding plates within his throat to break up prey.

G / Ice fish. *Waters of the Antarctic, where the ice fish lives, are rich in food for active and passive feeders.*

H / Shark. *The shark is undoubtedly an active feeder. Stimulated by blood or the sounds of a struggling fish, he may attack in a rush of fury.*

I / Orca. *The orca has been given an undeserved name—the "killer whale." He is no more a killer than any other animal who must actively capture a meal.*

J / Barracuda. *The barracuda is an extremely fast swimmer—using lightning darts and ferocious teeth to capture prey.*

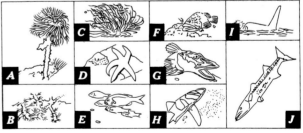

Safety in Numbers

The sea is the mother of life and a prodigal mother she is—or perhaps the word is profligate. The jet-propelled *Loligo* squids on the opposite page are only ten inches long. They live two or three years in open waters, often in small groups. Then they gather for a mating orgy that lasts three or four days. The crowd of frenzied mating creatures is so thick that we have been able to estimate a count over a billion squid in a volume only 400 feet in diameter and 100 feet deep. Such are the astronomical dimensions of reproduction in the sea. Such is the price of species success in the sea. Hake lay up to 1 million eggs at a time; haddock up to 3 million; cod to 9 million. In 1932 careful observations off the southern shores of Cape Cod indicated that in this one spawning area 1 billion mackerel produced 64 trillion eggs. It was calculated that of each million eggs laid only 1-10 fish survived to maturity; in other words, the mortality rate ranged between 99.98 per cent and 99.99 per cent.

This is good enough. The mackerel is one of the most successful species in the ocean. In fact, if more were to survive the population would grow in size and possibly upset the delicate balance of nature. An abundance of mackerel may deplete their food resources and in turn lead to mass starvations—thus perpetuating the imbalance but in an opposite direction. It seems that many species produce millions of small eggs and merely cast them to the currents to fend for themselves. In contrast, other coastal fish remain to protect and nurture a few larger eggs.

In the sea there is no best way to live — whatever method is successful will be retained. The laws of nature have been set by life itself. The individual as individual is not important. In many species he dies as soon as he breeds. By feeding his own small input into the "natural selection" computer he has accomplished all he is able to do: if he has survived long enough to breed he is a success; if he has not he is a failure; in neither case is he remembered. Most fishes reproduce by means of eggs which are fertilized after they have been laid. But the oceanic reproduction process is anything but simple. In some species functional hermaphrodites

> "By feeding his own small input into the natural selection computer the individual has accomplished all he is able to do. If he has survived long enough to breed he is a success; if he has not he is a failure; in neither case is he remembered."

are found — individuals that can produce both ripe sperm and eggs. In others the sex changes from male to female during the growth process. Others fertilize themselves. Fishes cannot afford the regular gestation schedules of most land animals. In some species the female accommodates her spawning time to water temperature. Others store sperm in their own bodies. Others bear larvae. With the ovoviviparous sharks the embryos are released from their eggs but continue to live inside the mother's hinder parts. In the egg-laying species there are innumerable methods of protecting the eggs—by fastening them down, shielding them with foams and membranes, carrying them in the mouth (of the male) or in pouches. But in the course of their evolution fishes have learned that real safety lies in numbers.

A closeup of a squid mating orgy. "*The crowd of frenzied creatures is so thick that we have been able to estimate over a billion squid in a volume only 400 feet in diameter and 100 feet deep.*"

The Spawning Salmon

"I know of no more melodramatic account of the climax in the life-cycle of the salmon than this story of a fourteen-inch male silver salmon in Humboldt County, California, related by Robert T. Orr in *Animals In Migration*.

" 'This individual, as well as a number of others, was removed from the hatchery at one year of age and released in another nearby coastal stream. The following year at spawning time this male was found back in the hatchery tank. To get there it had to come in from the sea and up the home stream for five and a half miles, follow a rivulet across porous ground, and travel under U.S. Highway 101 through a one and a half foot culvert and into a storm sewer. From there it was necessary to jump into another smaller culvert and up eighty feet to a flume. Once in the flume the only way for the salmon to get to the rearing tank was through a four-inch drain pipe that had a ninety-degree bend in it. The pipe rose vertically two and a half feet from the bend to enter the tank. The outlet also had a screen cap covering the top and was surrounded by a square box of wire netting. The square was open at the top and only slightly higher than the water level in the tank. This fish had made this remarkable journey through the culverts and pipe and had knocked off the screen cap to the drain and leaped over the wire netting surrounding the drain. It was aptly christened *Indomitable*. Further investigation revealed that 72 other salmon which had taken the same course had become stranded in the flume.'

"What makes *Indomitable's* story terrifying, as well as other salmons' stories, is that his pilgrimage was not just to sexual maturity and fulfilment but also to death. When the mature salmon reaches his rearing-grounds, perhaps a tiny pond high up an Alaskan river, he is not then given the reposeful retirement we might think he has earned. Far from it. Bruised and torn by rocks and predators as he is, without food from the moment he has entered fresh water, his body convulses again and again in the orgasmic contractions of procreation — the female digging a hole in the rough bottom for her eggs, the male fecundating the eggs with sperm. Having completed this function there is nothing more for salmon in life. They begin to age at a fantastic rate. And in a few days, utterly spent, they simply drift to the shores of their little home and die."

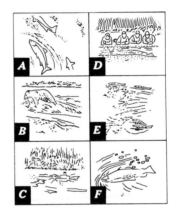

A / Rapids. *Jumping water hazards to get to their mating grounds.*

B / Dangers. *Just one of the perils the salmon faces during his journey is hungry brown bears.*

C / Going upstream. *The salmon homes in on the home scent.*

D / Onlookers. *Captain Cousteau and his crew clocking the salmon.*

E / Mating ground. *The surviving salmon swimming placidly.*

F / Death. *The salmon's purpose over and energy depleted, he dies.*

Home In the Harem

You are looking in on a small part of a colony of northern fur seals deployed over one of their breeding grounds. The central institutions of seal society are the harems, organized in late spring by mature bulls, five years old or older, who climb ashore at the rookeries after ten months of "batching" it in the open seas. There is a good deal of shoving and heaving, and some serious fighting, before the subadult males and the aging ones have been chased out of the main camp and each breeding bull has staked out a private territorial claim. At this point the females swim ashore—most of them in the last stages of their pregnancies from the previous season. Each female finds her bull and moves into his harem. How large a harem depends on many factors, but it can be sizeable; a big male northern fur seal can control as many as 100 cows. The pups are born, and within a few weeks the adult seals have

> "At the height of the season a male northern fur seal can mate three times an hour and can maintain a rate of one act of love an hour for periods lasting up to three weeks."

mated again after a courtship of perhaps only a few moments. (The bull seal is no refined lover.) Cows tend to be recruited into harems at about five years of age, but it is known that a few four-year-olds breed outside the colonies—adolescents have created

problems since the beginning of time. Twenty to thirty days after the mating period the harems break up and the bulls depart, leaving behind their females and suckling pups, not to see them again until the next spring. One of the reasons for such a "sexist" arrangement may lie in the size-discrepancy between male and female seals. Very much bigger and stronger than their mates, the males may constitute a danger. The males pay for these prerogatives. It is only rarely that a male seal attains the age of twenty-five, while females can live as much as ten years longer. Still, it is all very similar to an Arabian Nights establishment—which is why the appellation "harem" was applied to the seal colony in the first place. At the height of the season a male northern fur seal can mate three times an hour, and can main-

tain a rate of one act of love an hour for periods lasting up to three weeks. Fast and big and intelligent, seals take pretty good care of themselves in the open sea. Only the faster and bigger and more intelligent killer whale is too much for them. But in their summer rookeries—overwhelmed with the pressing business of establishing territories, fighting off interlopers and eager bachelors, breeding, mating, scolding, suckling, and preparing for the long winter ahead—the seal is absolutely defenseless against the clubs and guns of the human seal-hunter. So those which carried valuable furs on their backs were all but exterminated from the planet. Perhaps man has stopped just in time, for with a few tragic exceptions the seal family —including walruses, sea-lions, and elephant seals—now seem to be fairly secure.

Attack and Defense

A / Shark. *Rows of sharp teeth line the mouth of the shark.*

B / Sea snake. *Its venom is lethal, but a sea snake has such a small mouth it can bite a diver only in places like the fold of skin between the thumb and index finger.*

C / Walrus. *The sturdy tusks of the walrus grow up to two feet in length.*

D / Blue crab. *This crustacean uses its oversized claws for courtship gestures as well as defense.*

E / Sea urchin. *The sharp, brittle spines of the sea urchin present a formidable defense.*

F / Surgeonfish. *Knife-like spines are found on each side of the surgeonfish near the tail.*

G / Orca. *The "killer whale," as it is mistakenly known, attacks with its large, sharp teeth.*

H / Moray eel. *The razor-sharp teeth in the mouth of the moray eel give this fish a menacing look.*

I / Lionfish. *The fearless lionfish is equipped with a row of 18 needle-like dorsal spines.*

J / Tridacna clam. *The heavy shells (bivalves) of the tridacna clam clamp firmly together.*

The Art of Motion

The dolphin below seems to be streaking through the sea with the greatest of ease, but he is achieving this whoosh by virtue of a number of complex adaptations to a difficult medium. Water is about 800 times denser than air; that is, a glass of water weighs 800 times more than a glass of air. Also, water is approximately 50 times more viscous than air, putting up a resistance to flow nearly 50 times that of air. These two factors, density and viscosity, affect the buoyancy, shape, size, and efficiency of propulsion of animals inhabiting the undersea environment. Neutral buoyancy solves the lift problem in most fishes. To rise or descend they need only extend their forefins at the proper angle, as submarines do with their diving planes. How fast they swim through

the water depends on their shape as it meets the water as well as their basic muscle power.

The water flow over the body of a moving fish can be smooth (laminar), irregular (turbulent), or a mixture of the two. Laminar flow produces least drag and turbulent flow most. Naturally, streamlining promotes laminar flow. Any structures projecting from the body tend to produce turbulence, so fins and flippers are streamlined, and in rapid swimmers they can be folded against the body. Turbulence may develop if the body surface is irregular, so the skin of most active swimmers is smooth and covered with a slimy mucous secretion. It all works, although it is difficult to measure the maximum velocity of the ocean's creatures, partly because a powerful sudden spurt after prey or in avoidance of capture is more important to most fishes than a high sustained running speed.

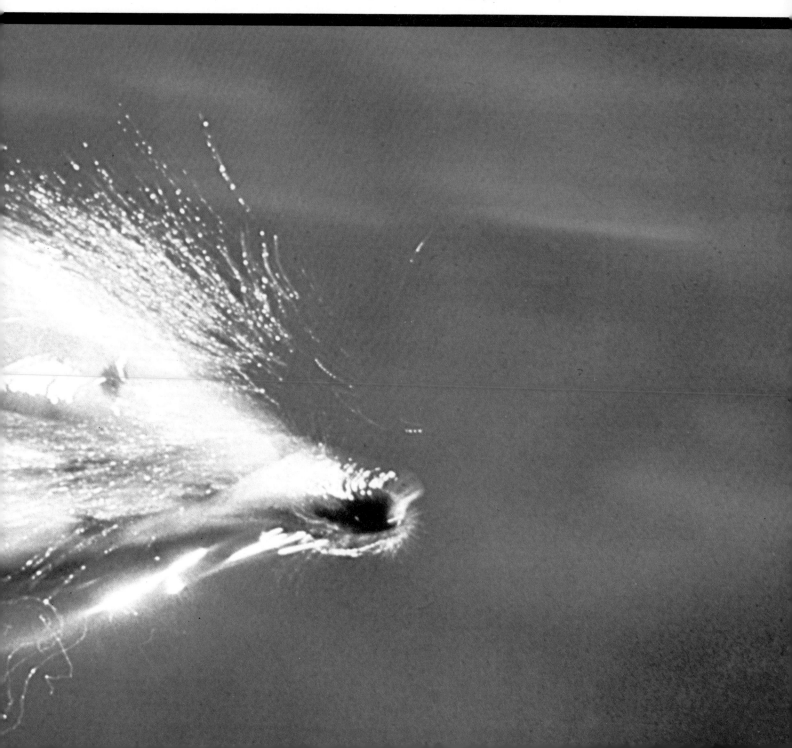

Fit for Travel

A / Jellyfish. The jellyfish unhurriedly travels about his domain, drifting with the currents.

B / Sea turtle. These sea turtles have flattened shells, streamlining them for swimming.

C / Dolphin. The streamlined dolphins shown here are ingeniously riding the bow wave of Calypso.

▲ A   ▼ B   ▲ C

D/Manta ray. This huge relative of the shark is relatively gentle. It flies like a bird about its submarine world.

E/Killer whale. The sleek streamlined body of the Orca assists this highly intelligent mammal in its life in the sea. Broad horizontal flukes, like those found on other whales, propel this superior dolphin.

Primitive and Modern Fishing

Early man hunted fish for the same reason fish pursue each other, in order to eat. Fish rarely kill more than they can consume, and in primitive societies fishermen, like fish, took from the sea only the quantity of food needed to feed themselves, their families, and the tribes or other social groupings for which they were responsible. That unwit-

> "Man has become by far the greatest predator of all time."

ting conservation of the riches of the sea was dictated partly by tradition; it was also a consequence of the limited gear the fishermen had to rely on.

About a hundred years ago, Thomas Henry Huxley, the English biologist, could say with confidence: "I believe probably all the great sea fisheries are inexhaustible; that is to say that nothing we do seriously affects the numbers of fish." Huxley, of course, did not foresee the technological advances that would be made and applied within a century to fishing as well as to every other aspect of man's activities. Sadly, Huxley's prediction has proved wrong. Today's fishing fleets consist of entire flotillas equipped with electronic devices and include floating ships that process the catch at sea. Planes are used as spotters to locate schools of fish. Radio telephones direct the boats to the fish. Radar and echo sounders find schools that cannot be seen from the air or on the water's surface. Moreover, oceanic sciences have determined the conditions of salinity and temperatures required for various species to thrive. And today, thermometers and salinometers are used by fishing fleets. These ultramodern fishing fleets can stay in the open water almost indefinitely, sweeping the sea of much

of its life. The herring population in the Atlantic, the most heavily fished area of the world, is decreasing. Haddock may have been wiped out. Similarly, modern whaling methods have all but eliminated several types of whale from the face of the earth. Helicopters and swift engine-driven catcher boats with sounding equipment spot the whales. Explosive harpoons fired from guns mounted on the catchers find their mark.

Man has become by far the greatest predator of all time. As populations mount and land-grown food supplies are unable to feed the growing numbers of hungry, man is turning more and more to the sea for his food. On land man has slowly learned to conserve the soil lest it stop producing crops. But on the ocean, man is a hunter only. He takes but returns little. If the bounty of the sea is not to be exhausted, man must learn to farm it as he farms the land, by sowing as well as reaping.

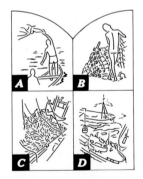

A / Net hurling. *A simple fishing method for feeding a small community.*

B / Drying the nets. *Satisfied with the day's catch, fishermen spread their nets to dry.*

C / Commercial fishing. *Typical day's catch aboard a modern fishing vessel.*

D / Surrounding the catch. *Modern fishermen prepare to haul in their nets.*

Chapter V. Feeding Information Into the Computer

Man perceives the world through his senses. His brain receives a steady flow of information about his surroundings in the form of elaborate messages from his sense organs. On the basis of those sensory signals plus memory, his brain selects and produces appropriate responses. All land and marine animals possessing brains and nervous systems learn about the external world in like fashion. The more senses such organisms have and the more sensitive those senses have developed, the more efficiently and creatively the creatures will perform. Because the brain processes incoming information and responds to it, this most

> "We know how fish see and hear and smell. But that information does not really tell us how the world looks or feels to them, nor does it explain some behavior that depends on abilities we do not ourselves possess."

complex of all body structures is often compared to a computer. But although computers can perform some mathematical and analytical functions more speedily than the brain, the living brain is a far more complicated mechanism than any man-made machine. It has a greater capacity for storing memory and handling information than it actually uses. As a living organism-computer the brain also feeds on a much·wider spectrum of input: hormones and other trigger-substances in the bloodstream, inherited and unconscious memories and precognitions of all kinds, exquisitely discriminate messages from the skin and the senses.

The cut-off point on what an organism perceives comes generally not from deficiencies in the brain but from physical limitations on the senses themselves. Man, for example, can only hear sounds within a given distance and within a given range of frequencies. The dolphin can hear more acutely at greater distance and over a wider range of frequencies. Similarly, man's sense of smell is less discriminating than that of a dog or a shark and his vision is less keen than that of a hawk. He may also be less sensitive to touch than the lowly featherduster worm, which retracts into its tube when approached by any other animal.

Man has catalogued the senses of the rest of the animal world and tried to ascertain their full range, but he has necessarily done this in terms of his own sensory equipment. Thus we know whether and how well fish and the aquatic mammals can see and hear and smell, and how sensitive they are to taste and touch. But that information does not really tell us how the world looks and feels to other animals nor does it explain some animal behavior that depends on abilities we do not ourselves possess, such as the echolocation of bats and dolphins. We know that all animals, even the lowest ones that lack both a brain and a nerve network like the one-celled amoeba, still have some system for responding to external stimuli. Can we imagine animals with sensitivity to gravity, magnetic fields, and cosmic rays? Such senses probably exist and guide the migrating animals! Or animals than can communicate via telepathy outside existing channels?

The tropical crayfish relies heavily on his senses of touch and vision for protection and food. His extended feelers, sensitive to touch and chemical variations in his surroundings, give greater ability to detect objects without his needlessly exposing himself to possible danger. The crayfish's compound eyes provide his brain with a composite image of his world.

The Senses

Animals must obtain information about their environment to live successfully in the sea. Specialized sensors are utilized by the higher animals while lower animals perceive their environment on a basic cellular level. Simple organisms probably respond, in one way or another, to everything we respond to—such as heat, cold, touch, chemicals (smell and taste), light, and sound, the difference being lower sensitivity.

A / Sponges. Like the fish nearby, these sponges are sensitive to light and various chemical stimuli. For example, the release of sexual products is stimulated by chemicals from adjacent sponges and the development of these gametes is controlled by day length.

B / Catfish. Barbels, small feelers under the mouth of a catfish, give it the ability to taste food before eating it.

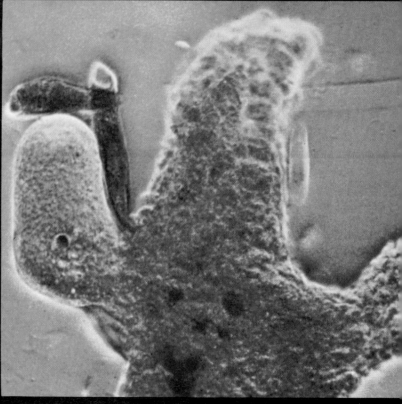

C / Octopus. *The octopus relies primarily on vision. Eyes similar to man's aid him in feeding, camouflage, and locating a mate.*

D / Sea gull. *Marine birds such as this gull have keen eyesight which allows them to locate food from great distances.*

E / Feather duster. *This feather duster worm can without elaborate sensors perceive the pressure waves from a diver's moving hand or the shadow of a passing fish.*

F / Amoeba. *An amoeba can, without any specialized sense receptors, chemically locate and capture a meal, withdraw quickly from a bright light, and react positively or negatively to temperatures.*

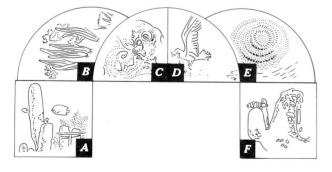

Comparing Senses of Man and Dolphin

Man is a visual animal. The dolphin is an acoustic animal. Each has what he needs most in his own world. From a hilltop or promontory, on a clear day, the normal man can see as far as 100 miles. But on the "clearest day" in the sea the eye cannot reach much farther than 100 feet. Even crystal-clear water is actually a fog so far as the organs of sight are concerned. In returning to the sea the dolphin had to adapt to this suggesting that the entire structure could be involved in responding to sound. In practicing echolocation the dolphin emits a fascinating range of noises—clicks, tone signals, whistles—both for bouncing ahead of him or communicating with his fellows. His hearing range is very much larger than man's—going up to 150,000 cycles to man's top limit of about 12,000. Of the dolphin's other senses probably the visual is next in import-

undersea limitation. He was able to compensate thanks to the hypertrophy of his auditory faculty, modifying greatly the structures of the internal and external ears in order to receive and interpret a wide range of waterborne vibrations. All this not only vastly improved the animal's hearing

> "From a hilltop, on a clear day, a man can see 100 miles.
> But on the clearest days in the sea the eye cannot reach much farther than 100 feet."

underwater but refined it for the purposes of echolocation—the sonar-like process with which the dolphin homes in on objects as distant as a mile away that he wants to inspect—perhaps to eat. In a sense the dolphin's whole skull may be an "ear," because his ear bones are connected to it in a fashion

ance, though he cannot match man in this faculty. Dolphins in marinelands see well enough to leap out of the water for the fish their trainers dangle before them. He must also experience some sensations of his skin, otherwise he would not make love as he does, but of course he has no hands—and therefore little sense of touch. As the dolphin possesses no olfactory nerves, we must assume that he has no sense of smell, but it is possible that some hitherto unidentified area of cranial tissue may deliver to his brain some of the aspects of smell. Nor does the dolphin seem to possess taste buds or salivary glands. What he does have is a brain comparable to that of man.

Second sight. *Although humans and dolphins can see about the same distance underwater, the dolphin has an extra sense, a "sonar," which permits him to perceive events precisely and over much greater distances.*

A Sixth Sense—the Lateral Line

Fishes thrive in a three-dimensional space, in constant peril of predators, at least twice a day on the prowl for their own nourishment. Many species swim in schools of thousands of like-sized comrades amidst the crashing surf or storms or through the intricacies of a reef, in the depths where the sun reaches weakly at midday or at night when the ocean is entirely dark. Altogether:— conditions which reduce the traffic problems at a modern urban airport such as New York's Kennedy to as simple an operation as walking across the street.

How do they manage? How do fishes hold onto their order and orientation? No one knows exactly. But the solution probably lies largely in what we call the "sixth sense" of fishes—a longitudinal system of canals and openings which starts over the head of most fishes and runs under the skin along the body to the tail. This "lateral-line" organ almost certainly has two jobs: that of perceiving distant pressure disturbances as well as changes of flow patterns close to the fish. It is a sort of "pressure ear," similar in anatomical structure to the semicircular canals of the "auditory ear."

"A moment's thought tells us how essential such an organ must be. Many times I have been diving peacefully near a school of fish when, all of a sudden, the formation breaks and the individual fishes scatter into every sort of crevice and hideaway. What is the matter? I have heard nothing—seen nothing. Two or three moments pass before I spot, in the distance, an oncoming group of dolphins or sharks. The smaller fishes 'heard' the pressure waves created by the silent predators long before they came within sight-range—and cleared out of the way."

The fish needs a special organ for this. One of the laws of acoustics is that the lower the frequency the larger must be the receiver. Pressure waves possess a very low frequency. The fish ear it not nearly big enough to receive such low frequencies, so to obtain information about distant vibrations fish rely upon what amounts to an accessory acoustic organ. The lateral line is all the more useful to him because in sea water the lower frequencies yield the better transmissions. As a fish swims forward he pushes

> "Nothing is more difficult to imagine than the input of a sense you yourself cannot experience."

ahead of him a small pressure wave which then flows around his body. By "listening" to the reflection of this pressure wave he can detect objects near him. He can, for instance, avoid bumping into the glass walls of his aquarium. If caught in rough water, he can steer to quieter areas.

Beyond these generalizations we cannot now go—partly because no land animal possesses any such organ as the lateral line (neither do any of the reptiles or mammals like whales or dolphins which have returned to the sea), and nothing is more difficult to imagine than the input of a sense you yourself cannot experience. What we do know is that all fishes have pretty good ears, most have lateral-line organs and eyes—and in pursuing their life in the sea, the success of which depends largely on the sharpness of their "traffic control," they must blend messages from these three senses into an efficient "picture."

A jack with a conspicuous lateral-line system. This fish is a speed merchant. Note the highly developed ventral, or bottom, fin, which acts as a sort of flexible keel to give its owner extra driving power.

Chapter VI. The Computer

The human brain is in constant communication with all parts of the body. It controls breathing, digestion, and rate of heartbeat. It regulates the chemical changes that take place ceaselessly within the organs. It receives messages from the senses and directs and coordinates muscular activity. It houses memory and is the seat of the emotions. It is the site of the highest of mental functions —thinking, judgment, imagination, creativity, and consciousness.

These many activities are carried out by nerve cells called neurons, which are like miniature FM radio units that both receive and transmit complex messages, so that each neuron is many times more elaborate and efficient than one computer circuit. The human brain has about 15 billion of these cells, many times the number of on-off switches in even the biggest contemporary computers. Each neuron may be in communication with as many as 270,000 others. The total number of possible connections between neurons is all but incalculable.

The dolphin in the sea has a brain whose complexity and number of nerve cells compare with that of man and is many times superior to the brain of the higher apes. Both dolphins and apes have inhabited the earth longer than man, yet neither has ever produced a civilization. Why not, one may ask? Civilization can only develop if past and present learning, history, artifacts, and traditions of a group of individuals can be preserved and stored in such form that it can be communicated from one generation to the next even as it grows larger.

For a civilization to develop at least five factors are evidently required, and man alone of all creatures on the evolutionary ladder has been the beneficiary of all of them simultaneously. There must first be intelligence. There must be the means of using and making tools, which makes writing or electronic storage possible, and thus enables communication to overlap generations. There must be a flexible means of communication, a language. There must be a sufficiently long lifespan for the acquisition and transfer of new knowledge. And the basic needs, such as those for food and shelter, must be easily taken care of so that energies can be directed elsewhere. Civilization can develop only when survival is no longer of paramount concern.

Dolphins have intelligence. They live long enough to acquire and store a considerable body of knowledge about their environment. Their basic needs are taken care of with time to spare. They have a means of communication. Through a repertory of various whistling sounds, dolphins appear to talk to each other in much the same fashion as those isolated human groups that have developed a

> "No matter how intelligent dolphins may be,
> without the help of man they will never advance to
> a higher level of performance."

"whistle" language. But dolphin messages, being aural, dissolve in the water through which they are sent just as human whistle-languages die in the air. Even if they had developed an "oral tradition" system, dolphins have no hands to write with; their communication is limited to the present. No matter how intelligent these creatures may be, without the hypothetical help of man they will never advance to a higher level of performance.

The Basic Unit

The neuron is the basic unit of a nervous system—any nervous system, whether it be a worm's or a seal's. Its function is communication from one part of the body to another. Most protozoa (amoebas, etc.), being small, have no need for elaborate communication networks and thus do not possess nerve cells. But all groups of animals from anemones on up build upon the basic unit of internal communication—the neuron.

Essentially the neuron is composed of a cell body containing a nucleus and other vital organelles from which extend branching connectors allowing communication with other cells. Incoming impulses are received by dendrites and transmitted through the axon and its branching to the next cell, and so on. In contrast to simple animals which act without thinking, higher animals collect information by receptors, evaluate it, and act appropriately. Simply put, neurons communicate information to the brain where other neurons communicate internally among themselves to reach a "decision" which is directed via still other neurons to muscles for the proper response. The neuron transmits an impulse electrically along its surface. Some even use a specialized sheath cell to surround the axon which acts like the insulation of electrical wires.

It is amazing to think that learning, memory, and instinct all develop from varied associations of simple neurons. In an octopus which just learned to reject a crab or accept a shock, how do the axons adapt so as to stimulate dendrites other than those of "normal" channels? No one knows.

The neuron. *It is fairly certain that we will not in the near future unravel the mysteries of the simple neuron and its function in the complex process we call thinking.*

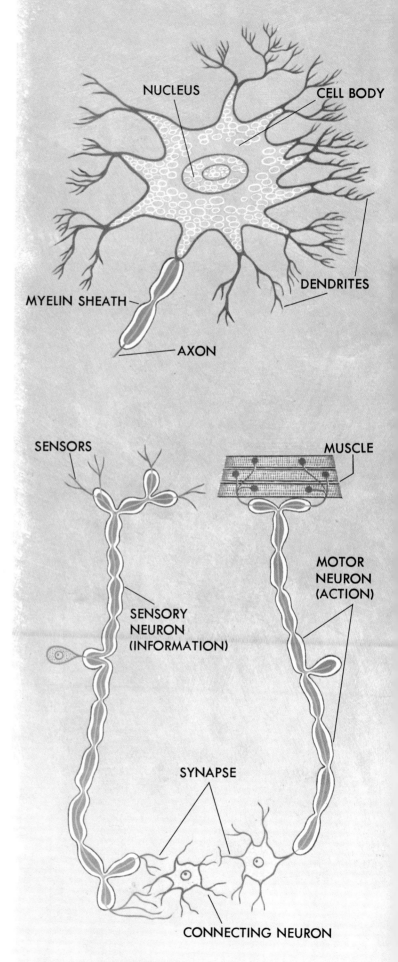

▲ A   ▲ C   ▲ B   ▲ D   ▶ E

Problem-Solving Octopus

Biologists consider the octopus the most intelligent of the invertebrates. The octopus is a mollusc and therefore distantly related to oysters, snails, and scallops and more closely related to the squid. The octopus brain is the largest and best developed of all invertebrate brains. It is divided into 14 lobes, each governing one or more functions. The largest governs sight and there are others for memory and reasoning.

A / Favorite food. The diver took a covered glass jar containing a lobster down to the ocean floor.

B / Tempting the octopus. The diver then placed the jar, with a hole in the cork to allow water to pass through, near the nest of the octopus.

C / On the lookout. Always searching for food, the octopus spots the lobster—not knowing, of course, that it is in a closed jar and therefore a challenge to get at.

D / Closing in. The octopus reached a tentacle out to catch the lobster and was confronted by an unusual puzzle—he couldn't get to his food. Irritated, the octopus tried to sweep the jar into its home, but it wouldn't fit. It was noticed by our team that the octopus became angry and anxious—he breathed harder and his heartbeat speeded up.

E / A meal well earned. But the octopus did not give up. After three hours of luck or logic, and trial and error, the cork was removed and the octopus claimed the prize.

The Potential for Learning

Intelligence, meaning the capacity to apply what one has already learned to new circumstances, is generally thought to be a function of brain size. As a rule, the larger the brain the greater the potential for intelligence. However, size alone does not guarantee superior intelligence. The elephant has a bigger brain than man but is not more intelligent.

Brain size is correlated with body size. As an animal increases in overall size, its nervous system must likewise increase in size or the animal will become unable to meet the challenges of its environment. One reason the dinosaur became extinct is that its body outgrew the ability of its nervous system to function efficiently. Given an adequate brain size and presupposing that animals continue to learn as long as they live, how much an animal learns will de-

> "The dolphin is more intelligent than any other marine creature. It is capable of creating—that is, of initiating new behavior."

pend on how long it lives. Man, as one of the longest-lived of animals, can, over a lifetime, learn more than most other creatures. And one thing he has already learned is how to increase his own lifespan. More than a century ago, in 1850, the average life expectancy in the United States was about 41 years. Today, largely through expanded medical knowledge of formerly fatal childhood diseases and the development of life-saving drugs, the average life expectancy has risen to 70 years.

The dolphin, having a large brain and living to about 30 years, acquires much information in a lifetime. Since dolphins make many kinds of noises—grunts, calls, clicks, squeaks, barks, and whistles—some observers presume that these sounds are the means by which the animals communicate with each other. By communication the observers do not mean calls for help or cries of pleasure, which are emitted by many animals, but the actual transfer of information through "language." Doubt is cast, however, on whether dolphins have a true language and even on their general level of intelligence by the fact that the creatures are caught over and over again in the same regions of the world in the nets fishermen put out for tuna. One would suppose that if dolphins truly had both language and high intelligence one of them would have warned the others long ago and all dolphins would have learned to avoid these dangerous waters.

In any case, the dolphin is surely more intelligent than any other marine creature. Recent laboratory experiments shed some light on this intelligence. Some researchers in Hawaii have taught dolphins to create —that is, to initiate new behavior. The experiment began by teaching a dolphin a trick and rewarding him for it. Upon returning a day later the trainer would not give a reward for that same trick. Eventually the dolphin would give up and start to play, whereupon the first new act in his play would be rewarded. After a few days the dolphin would begin each session not with an old trick but with a new behavior. It was concluded that the dolphin understood that he must "think up" something new to please the trainer and earn his prize.

Lifespan and civilization: The longer the expected lifespan the more information the animal can gather and use. In turn, as in the case of man, that knowledge has helped him increase his longevity.

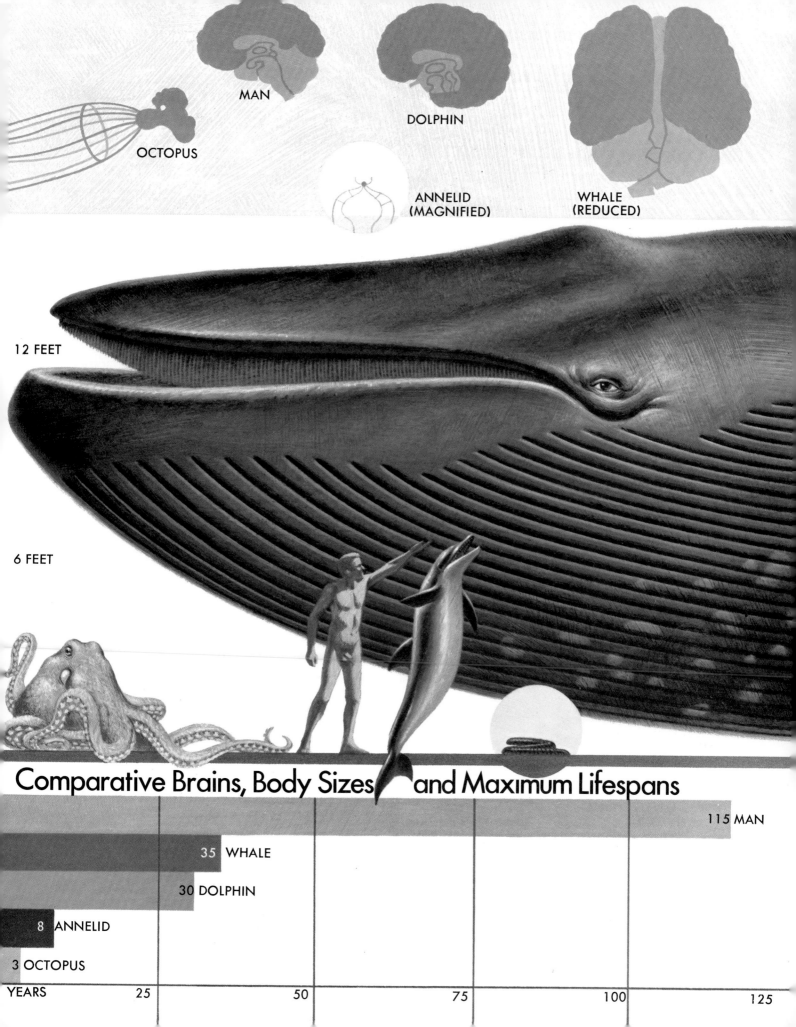

OCTOPUS

MAN

DOLPHIN

ANNELID
(MAGNIFIED)

WHALE
(REDUCED)

12 FEET

6 FEET

Comparative Brains, Body Sizes and Maximum Lifespans

115 MAN

35 WHALE

30 DOLPHIN

8 ANNELID

3 OCTOPUS

YEARS   25   50   75   100   125

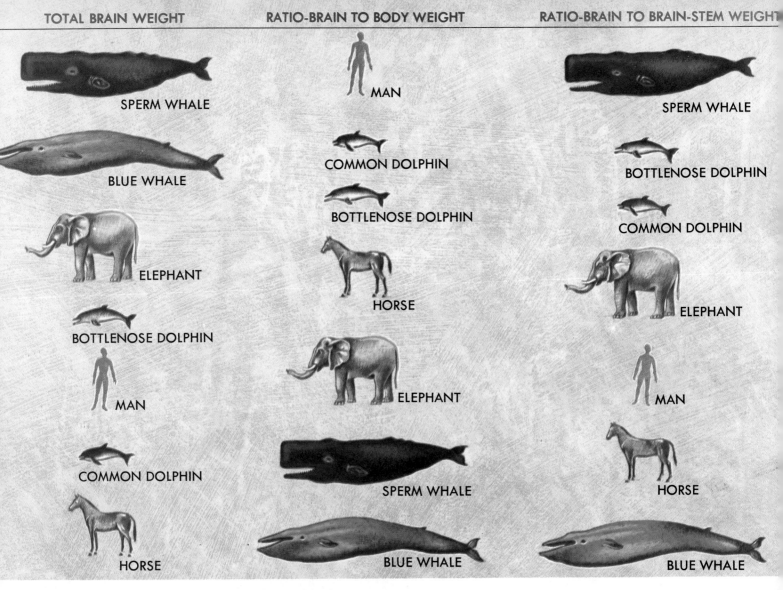

TOTAL BRAIN WEIGHT
SPERM WHALE
BLUE WHALE
ELEPHANT
BOTTLENOSE DOLPHIN
MAN
COMMON DOLPHIN
HORSE

RATIO-BRAIN TO BODY WEIGHT
MAN
COMMON DOLPHIN
BOTTLENOSE DOLPHIN
HORSE
ELEPHANT
SPERM WHALE
BLUE WHALE

RATIO-BRAIN TO BRAIN-STEM WEIGHT
SPERM WHALE
BOTTLENOSE DOLPHIN
COMMON DOLPHIN
ELEPHANT
MAN
HORSE
BLUE WHALE

Ranking mammals in three categories of brain ratios: Every order of animals, whatever the brain size of its individuals, has evolved its own highest species. The animals on the page opposite, each at the top of their respective evolutionary lines, usually exhibit some form of physical or sensory prowess, some kind of skill, and, often, a capacity for learn-

ing. In addition, they appear to be able to develop alternative reponses to new situations, thereby gaining a distinct survival advantage. We know the whale, the killer whale (bottom right), the octopus, and the dolphin to show what we call intelligence in captivity. They doubtless exhibit the same trait in their natural state.

Brain-Body Relationships

A / Dolphin. The dolphin's brain in relation to its body weight is only slightly less than a human's.

B / Sperm whale. The brain of the sperm whale, on the other hand, although it is heaviest in actual weight, ranks well below elephants, horses, dolphins, and men when measured against body weight.

C / Octopus. The most intelligent of invertebrates, the octopus has a brain which is relatively large, reflecting its surprising capabilities.

D / Orca. The orca is a highly intelligent animal

whose brain size is appropriately large when measured against the weight of its body.

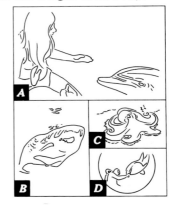

Chapter VII. Behavior Patterns

Life in the sea is so ancient that an almost infinite variety of behavior patterns have developed. Time is the key to the evolution of the many forms of feeding, moving about, mating, attacking, and self-protection that exist in ocean life. The rarest of all behavior is the self-destructive act of suicide.

Among the spectacular occurrences that take place annually in the sea, and must have or once have had survival value, are the migrations of large numbers of animals. Ocean migrations must orginally have been motivated by the search for food or a safe place in which to spawn. Now they often end in death and seem inexplicable, the animal counterpart of what in humans would be called compulsive behavior. Thus the freshwater eels of North America and Europe journey thousands of miles to the

> "The rarest of all behavior is the self-destructive act of suicide."

Sargasso Sea, a region in the Atlantic Ocean near Bermuda that is festooned with patches of floating seaweed, to spawn and die. The newly hatched fry float on Atlantic currents back to their respective coastal streams where they mature in fresh waters.

The 4000-Mile Meander

Each year during the months of December and January the California gray whale migrates southward along the western shores of North America, from the Arctic Ocean and Bering Sea to the tip of Baja California, to calve. Along the route the whales swim in relatively shallow water, generally between six and 100 fathoms (36-600 feet), from an area near Vancouver, British Columbia, past the west coast of the United States, to the desolate Mexican shores. The route of the northern portion of their journey from the frigid Arctic Ocean to Vancouver is not certain. Some scientists believe that the whales follow the 100 fathom curve, staying relatively close to shore. Others feel that the whales cut diagonally across the Gulf of Alaska from the Bering Sea to Vancouver.

During the course of this journey the whales eat practically no food, after having gorged themselves for long summer days in the krill-rich waters of the summer and fall Arctic Ocean.

Mother, child, and guest. In the large photograph, we see two California gray whales in the midst of their journey from frigid Arctic waters to the hospitable warm waters of Scammon's Lagoon. In the inset photo a diver swims with a baby gray whale while its mother looks on. The baby can expect to grow to a maximum of 45 feet.

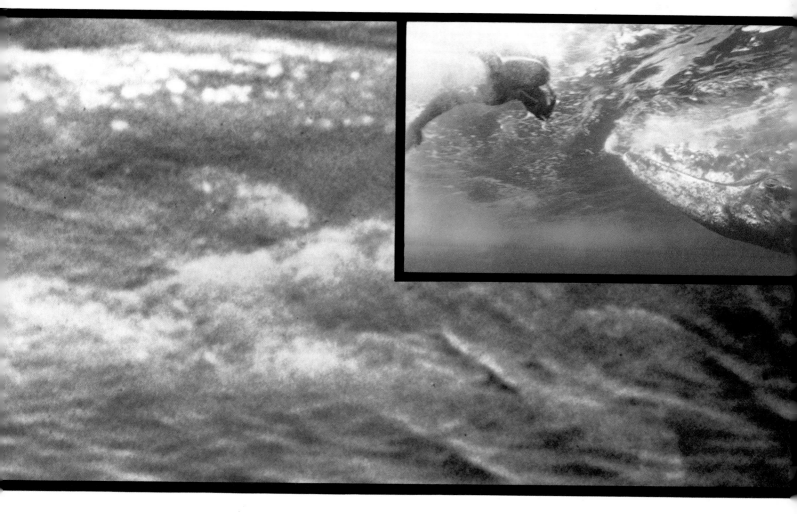

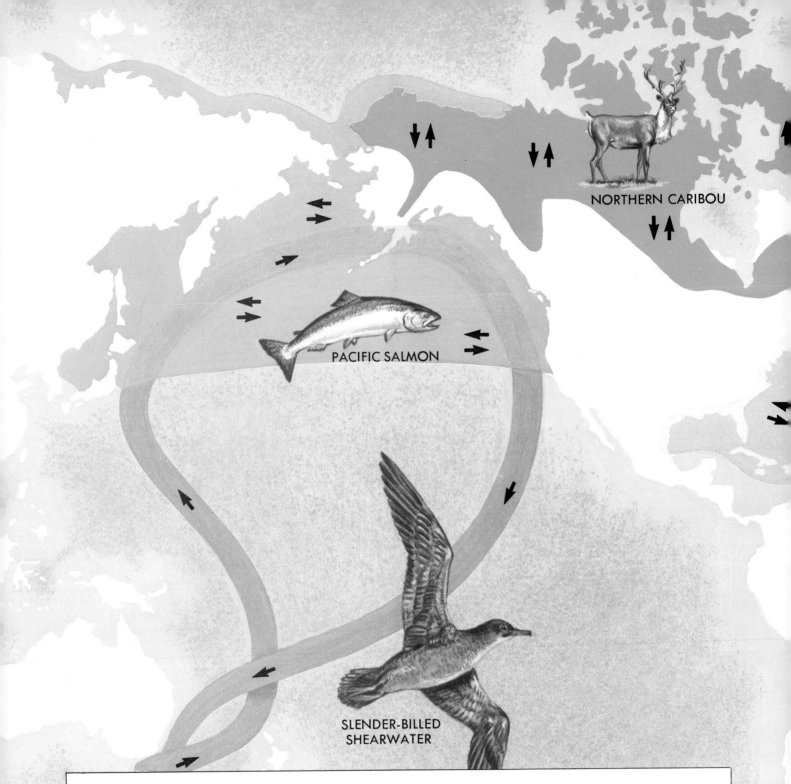

NORTHERN CARIBOU

PACIFIC SALMON

SLENDER-BILLED
SHEARWATER

Six Migrants

Many species of animals undertake seasonal or other cyclical migrations for a variety of reasons. Probably the most familiar of these are the seasonal migrations of birds seeking a more hospitable climate. But other animals migrate as part of their life cycle, for breeding or in search of appropriate nourishment, and still others seek different environments for as yet unexplained reasons. For instance, the American golden plover travels over a complex route from the Canadian Arctic to Southern South America and back for an annual round-trip distance of more than 16,000 miles.

The European Anguilla eel travels for a year without eating to mate in the Sargasso Sea.
The American Anguilla also mates in the Sargasso Sea.

Green turtles banded on Ascension Island were later captured near the Brazilian coast.

Herds of northern caribou migrate north and south following the seasonal growth of plants.

The Pacific salmon spends its adult life in the open ocean. It returns unerringly to the river of its birth.

Blue whales of the Southern Hemisphere converge on Antarctic waters in spring. In winter they scatter to the north.

The slender-billed shearwater migrates to the north and east in an exaggerated figure-eight flight pattern.

EUROPEAN ANGUILLA

GREEN TURTLES

BLUE WHALES

Ways Animals Adapt

A / Moorish idol. *The extended fins of the moorish idol provide stability, its shortened body maneuverability.*

B / Toadfish. *This slow-moving bottom-dweller is well camouflaged and hides in tin cans and other debris.*

C / Jellyfish. *Only 1 per cent of the jellyfish's body is the muscle tissue it needs to move through water in lazy jet propulsion.*

D / Grouper. *The enlarged head of the grouper acts as a powerful suction pump for feeding.*

E / Regal angelfish. *The extended snout of this spectacularly colored shallow-water fish is perfectly fitted to pluck small invertebrates from their hiding places.*

F / The spiny puffer. *The scales of this puffer, or porcupine fish, have evolved into spines. When the puffer feels threatened, it is able to inflate itself by inhaling either air or water. The resulting thorny ball lacks appeal to most predators.*

G / One spot butterfly fish. *Instead of camouflage for protection, these fish fool their predators. The deceptive eyespot directs an attacker away from the head.*

H / Wolf eel. *This ferocious looking fish has a mouth filled with fangs and molars.*

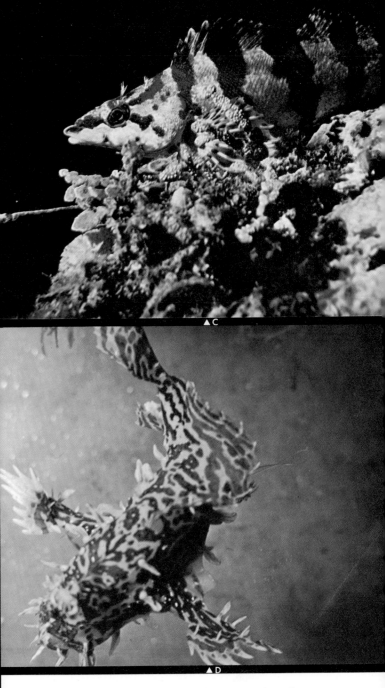

More Ways Animals Adapt

A / Sponges. The sponge lives very well without blood, a digestive system, a nervous system, or a means of locomotion.

B / Tridacna clam. The Tridacna clam of the South Pacific cultivates algae in "greenhouses" within its mantle. Sunlight enters through tiny spots in the mantle of the clam and its energy is converted by the chlorophyll of the algae, in combination with water and the carbon dioxide produced by the clam, into sugar and oxygen.

C / Klipfish. These fish bear their young alive.

▲ E

▲ F

▲ G   ▼ H

D / Sargassum fish. *The coloration and frond-like fins of the sargassum fish make it almost impossible to see.*

E / Jellyfish. *The pastel beauty of the jellyfish tentacles spell death for most small fish.*

F / Ocean sunfish. *These huge, slimy, scaleless fish have pitifully small brains. No one knows where they spend most of their lives.*

G / Sea horse. *A male sea horse incubating the eggs in its pouch deposited there by the female.*

H / Hammerhead shark. *The fleshy, T-shaped head of this shark may serve as a stabilizer.*

Symbiosis

The relationships we see here demonstrate one form of symbiosis, "mutualism," in which two dissimilar organisms live together or exchange services with each other for the benefit of both. The egrets that clean parasites off cattle show that the same behavior exists between land animals as it does in the sea. It is very likely that these associations improve the survival chances of the animals who indulge in them. The poisonous tentacles of anemones and some jellyfish are clearly a protection to the little fish who are immune to the toxin against predators who are not. Fish that patronize "cleaning stations" are freed of parasites and bacteria that might eventually cause them to die. In these pictures, each animal benefits from some characteristic of the animal shown with it.

The word "symbiosis" means "living together," an important part of the study of ecology, a much misused word these days. For ecology means the study of the interrelationships of organisms with each other and with the environment. Some of these relationships are between a plant and animal, like the algae that grow on the under side of the jellyfish, *Cassiopeia*, and provide it with an abundant supply of oxygen. In return, *Cassiopeia* lives upside down, providing the algae with a base on which to exist and a supply of needed sunlight. There are relationships between two kinds of plants, one alga growing on another, for example, or growing on one of the marine grasses. And there are the relationships between two animals that are mutually beneficial like those shown on the opposite page. Or like the crab that carries sea anemones around on its back. In that case the sea anemone benefits by being carried to new areas where there is more food. The crab, simultaneously, is camouflaged from potential predators.

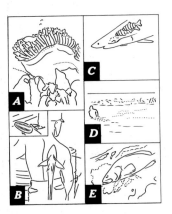

A / Anemone and clownfish. The clownfish, darting in and out of the sea anemone's tentacles, attracts food to the anemone, and even shares his meals with his host, which in return protects the clownfish with its poisonous tentacles.

B / Remora. This is an animal that literally hangs onto the shark—shown in the inset is the suction cup located on top of the remora's head. He hangs on, and from this vantage point is able to feed on scraps from the shark's meals. He earns his keep by cleaning parasitic crustaceans from the shark's skin.

C / Pilot fish. These fish generally swim directly in front of the shark's snout and seem to be of no practical use. Their numbers, however, may give the shark a certain status, indicating how successful he is at furnishing them with free food.

D / Fishermen and dolphins. The Imragen, fishermen of Mauritania, have a symbiotic relationship with dolphins. In winter, large schools of mullets come near the shore in very shallow water, where the dolphin's sonar gets confused by bottom and surface reflections. When the fishermen realize the mullets are there, they beat the water with heavy clubs to alert the dolphins swimming 20 to 30 miles offshore, who then rush to the shallows. The fishermen wade into the sea with their nets and the dolphins push the mullets into the nets, taking, of course, their share.

E / Cleaner and cleanee. The tiny yellow and black cleaning goby is here seen on a grouper. The goby sets up "cleaning stations" and open water fish wait their turns to be cleaned. The cleaners remove and consume parasites, diseased flesh, and bacteria from their customers, who grant them access to their gill slits and mouths.

Commensalism and Parasitism

There are other symbiotic relationships among marine creatures. In commensalism, two organisms live together. One of the commensals benefits from the relationship. The other is unaffected.

There are varying degrees of commensalism, some of which overlap with mutualistic relationships and some of which border on parasitism, yet another symbiotic relationship. In some cases a commensal lives attached to its ally. An example of this is a kind of hydroid that attaches to a fish's skin. Another form of commensalism is called "endoecism," in which a commensal shares a host's home burrow.

Another category of commensalism is called "inquilinism," which occurs when one organism takes shelter within another. The classic example is the pearl fish which dodges into the anal cavity of the sea cucumber when danger threatens. And there's the cardinal fish that lives in the shell of a living conch. And many of the bottom-dwelling gobies and blennies live in the pores of sponges. There are also several shrimps that live commensally in the pores of sponges. The sponge pumps in fresh supplies of food-laden water to the shrimp.

Inquilinism is the first step on the road to parasitism, a relationship in which one of the partners benefits to the detriment of the other who is the host. Perhaps the shrimp living in the sponge is depriving the sponge of needed food. But there are more clear-cut examples of parasitism. Some of the hydroids that live on a fish's skin send stolons or probing root-like organs through the skin and draw sustenance from the fish. And probably the most common fish parasite is the fish louse or copepod, a crustacean. Fish lice attach themselves to the skin or sometimes to the gills of the fish and suck blood.

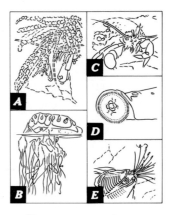

A / Prawn and anemone. *Anemones sometimes serve as hosts to the prawn which hides among the anemone's tentacles, finding protection there and sharing the anemone's catches.*

B / Portuguese man-of-war. *Living among the man-of-war's tentacles dangling under its mantle is the Nomeus. As long as it remains healthy and unwounded, the Nomeus is immune to the deadly sting of its host's tentacles, used to catch and paralyze anything coming within range. From this position the Nomeus shares its host's food. Should the little fish become disabled, however, it too is killed and consumed. Nomeus is not the only fish that finds the Portuguese man-of-war useful. Until they are large and strong enough to be independent, young amberjacks, cods, whitings, and other fry feed in the same planktonic meadows as the jellyfish.*

C / Hermit crab. *This crab resides in the shells of dead snails, periwinkles, etc., and changes shells in accordance with its rate of growth. It is interesting to note that in European waters some hermit crabs enjoy a symbiotic relationship with anemones, wherein the crab causes an anemone to attach to the crab's shell. The anemone eats the crab's food scraps, and the crab is protected by the anemone.*

D / Lamprey. *The lamprey is a parasitic animal which lives off other animals in a manner harmful to the host. It spends phases of its life in fresh and salt water and usually causes the death of the host. As an adult, the lamprey attaches itself to fish, often salmon or steelhead trout, by its suckerlike mouth and rasps a hole in the body wall of its victim through which blood and body fluid are consumed. The lamprey holds onto its swimming host until it has eaten its fill, or the host dies.*

E / Barnacles. *These are actually a species of shrimp which have built a shell around themselves. They grow not only on submerged structures but also on many forms of marine life. The barnacles affix themselves to a host, which provides a moving platform, keeping the barnacles in a generally food-rich environment.*

School for Survival

The schooling of many species of fish to be found today is 'one of those familiar phenomena we tend blandly to accept, forgetting the questions it raises. In what ways does schooling promote the survival of the species? No one knows for sure. But three guesses have been advanced. First, the school represents a concentration of breeding stock. The ocean is big, and the breeding opportunities of a species of widely dispersed individuals are obviously fewer than if those individuals bunch together in battalions. Then, schooling may be a way of minimizing the inroads of predators. This line of thought has been worked out mathematically. It leans on the fact that any one predator can consume only so much prey at one encounter. So if a school is kept

> **"When maintaining a close-alert in the sea 10,000 sets of sense organs are less likely than one set to overlook either an oncoming breakfast or an oncoming predator."**

at any size larger than this single-encounter bite the frequency of predator-encounters is minimized, individuals are concentrated, and those members of the school not taken in any one encounter are saved through the sacrifice of those that are.

Even those preyed upon are kept to a minimum by special markings or coloration. A disruptive visual effect is achieved through the use of bars, stripes, or silver coloration which makes it extremely difficult for a predator to focus on or pursue an individual among a maze of moving lines or shimmering color. Lastly, in schooling, aspects of group learning and group memory may be involved. When maintaining a close-alert in the sea 10,000 sets of sense organs are less likely than one set to overlook either an oncoming breakfast or an oncoming predator.

In any case, the schooling idea has proved its worth biologically—and, as typically happens in the ocean, has proliferated into countless varieties. Some fish school in tighter formations at feeding times. Some coalesce and then fan out, seemingly in random patterns. Some schools are spherical in shape—especially under the stress of attack. In a few cases the schooling instinct has been turned against itself; the barracuda, for example, has learned to "herd" schools of smaller fish for his own ends.

In the question of the school's mechanism —the way that the instantaneously-responsive, parade-ground discipline is maintained throughout the school—we now believe the eyes play the principal role. But many experiments indicate that more than sight is involved. It is a characteristic of schools that they consist not only of the members of one species but of individuals of that species which are like-sized. After disruptions, as for instance following the cross-through of another school, they are generally able to resume formation without difficulty. And fish can school in the dark. So we must suppose that other senses—perhaps the lateral-line organs responding to the many small vibrations set off by the movements of creatures through water— must be called upon to ensure the precision of the drill.

Traveling in schools. A typical behavior pattern of this species of herring. Less than 18 inches in length, this major food fish travels in schools of countless millions.

Chapter VIII. Interdependence

Nothing alive in the world lives alone. All living things are linked to other living things and to their nonliving surroundings in self-perpetuating cycles and networks called ecosystems. These systems, like the food chains of the world, progress from the relatively small and simple—the life in a given bay, for example—to the enormous and vastly complex such as the life of the oceans themselves. Even the sea, of course, does not exist alone. It takes material from the land, but also sprays the land hundreds of miles inland with aerosols rich in oligoele-

> "It has been calculated that the population of seals on the Pribilof Islands alone requires 3.5 billion tons of fish a year. Where does such a treasure of fish come from?"

ments and slowly but powerfully reshapes the continents. Like the planet itself, the watery sea is surrounded by a sea of air with which it conducts an unending exchange.

Man seldom thinks of himself as part of an ecosystem. Yet, for thousands of years, wherever he lived he was a part of the local ecosystem. A sudden discontinuity has been introduced by the industrial explosion. Man's influence has grown out of proportion. Which means that he must watch himself. In the pyramid of life in the sea, 1000 pounds of diatom fodder are required to support the growth of a single pound of commercial fish. At the center of this "oceanic food assembly line," in all open oceans, from the surface to about 4000 feet of depth, trillions and trillions of tons of creatures, minute plants and animals of all sizes, belonging to hundreds of thousands of species rise at night and sink during the daytime.

This universal "vertical migration" is the most important, the most majestic migration of them all. It is the "pulse of the oceans," governed by the sun. Early oceanographers were aware of this vertical phototropic beat thanks to plankton-net catches. But modern echosounders receive acoustic reverberations from the sinking and rising communities, which clearly show on the graphics of the sonars. The echos, resulting in the so-called "deep scattering layers," or DSLs, were often strong enough to be erroneously interpreted as the bottom, which led to a few cartographic mistakes. The DSL also has military applications, enabling submarines to hide under the DSL.

Today, the massive vertical migration has been thoroughly investigated, and has proven very complex. Nightly the DSL denizens feed on the "grass" of the sea—phytoplankton: tiny diatoms and dinoflagellates—and the zooplankton or microscopic animal life which feeds on phytoplankton. In turn the DSL animals form a food-supply for larger animals that ambush them.

The quantities of food involved in these marine round-ups beggar the imagination. It has been calculated that the population of Pribilof fur seals alone requires 3.5 billion tons of fish a year. Where does such a treasure of fish come from? It is now presumed that fur seals feed on one of the constituents of the DSL, the lantern fishes. Or consider the gigantic sperm whale, who can dive two-thirds of a mile beneath the surface for his dinner. He eats squids and abyssal fishes—but they eat the DSLs. And so it goes.

The pyramid of life. In the sea roughly 1000 pounds of plants will support 100 pounds of plant-eating animals which in turn will support ten pounds of meat-eating animals which in turn will provide one pound of human flesh.

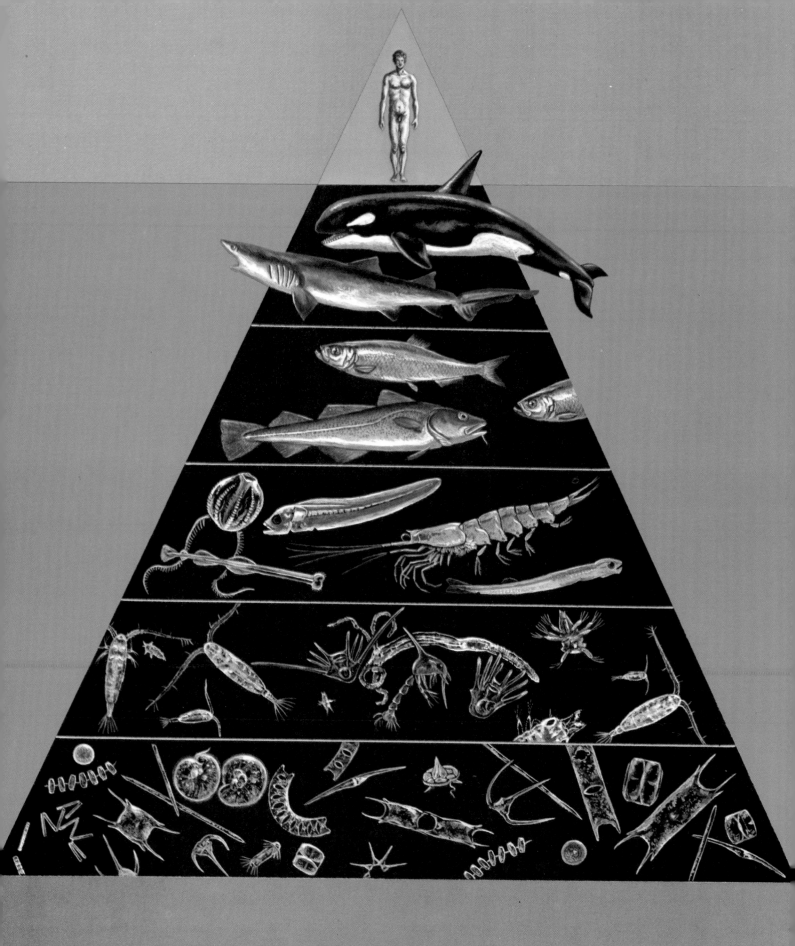

The Sea-Otter/Kelp/Sea Urchin Story

A striking example of how man can drastically alter the interdependence patterns of ocean life has occurred in our own time off the coasts of southern California. In the nineteenth century these waters teemed with that busy little carnivore the sea-otter, which ate sea-urchins, which ate kelp. When the fur trade all but exterminated the sea-otter the sea-urchin was left without a predator and responded by reproducing in vast numbers. These hordes of sea-urchins in turn reduced the rich kelpbeds to scanty patches. At this point man had replaced the original balance of nature with a new pattern, a prey-predator relationship between sea-urchin and kelp. As the kelp disappeared under this pressure the sea-urchins began to starve. With sea-urchin numbers down the kelp would make a comeback, upon which the sea-urchins would again increase, and so on—a cycle taking from ten to 12 years to repeat itself.

A next step came when sewage pollution caused an additional destruction of kelp—not because sewage kills kelp but because sewage feeds sea-urchins, once again booming their numbers. Before sewage entered the picture sea-urchins would starve when the kelp had been reduced below a certain level; now they lived on sewage. Some sewage outfalls in the Los Angeles area release over 300 million gallons of treated water per day. Near one of these in Palos Verdes there are literally millions of sea urchins. The area is a veritable desert populated almost exclusively by urchins. Were the sea-otter around in sufficient numbers, the kelpbeds—as important to this ecosystem as trees are to a forest—would in all probability still flourish. In their absence man must replace the otter in his role of urchin controller. Local dive clubs and graduate students periodically enter the area with hammer in hand and wipe out thousands of the purple marauders (or urchins). Their efforts are only a transient solution but in

> "Some sewage outfalls in the Los Angeles area release over 300 million gallons of treated water per day. Near one of these there are literally millions of sea urchins. The area is a veritable desert populated almost exclusively by sea urchins."

some cases allow replanted kelp plants to get a start. Of course, it would be more simple to have the sea-otter happily fulfilling his role in this ecosystem.

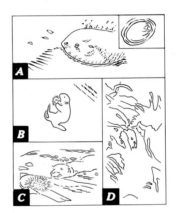

A / Abalone. *This mollusc, four to 12 inches across, is comparatively large. It is considered a delicacy for man as well as sea-otter. The inset photo shows the beautiful inside of its iridescent shell.*

B / Sea-otter. *This charming marine mammal, the smallest of sea mammals, is of course an airbreather. It rarely reaches 100 pounds in maturity and its tail constitutes one-fourth of its less than five-foot length.*

C / The meal. *Here a sea-otter, having brought a sea-urchin to the surface, prepares to enjoy it.*

D / Kelp forest. *Kelps are giant sea-plants which sometimes grow to over 100 feet. They serve as food for some sea creatures, cover for others.*

▲A ▼B

The Arctic Scene

A / Polar bears. *These huge animals are a source of both danger and food. There are now fewer than 10,000 in the world—mostly in Canada.*

B / Arctic fox. *This beautiful animal has one major natural enemy: the polar bear.*

C / Eskimo family. *A typical Eskimo family relies on polar bears, walruses, whales, caribou, and reindeer for their food and clothing. Armed with modern weapons, the Eskimo are beginning to endanger some of the species of their native regions.*

D / Penguins. *These birds, natives of the Antarctic, do not yet suffer the predations of man.*

E / Narwhal. *This whale with a single tusk is possibly one of the inspirations for the fabled unicorn.*

F / Gyrfalcon. *A predatory bird that lives off small animals and other birds.*

G / Seal. *An animal that has too frequently been useful to man for its lustrous coat.*

H / Wolf eel. *It is sometimes as long as nine months after birth before this eel begins to eat.*

▲C

▲F

▲D　▼E

▲G　▼H

At the Center—Calcium and Carbon

The beautiful reef on the opposite page has a story to tell—one of the most important in the book of the oceans. For like all other reefs this one is composed largely of calcium carbonate, and the cycling and recycling of calcium carbonate through the undersea world lie at its heart. On land calcium carbonate is found mainly in the form of limestone, deposits of which take up about 22 per cent of earth's land area. Rain and wind patiently gnaw away at these deposits, washing them into the sea in solution as calcium bicarbonate at the rate of 500 million metric tons each year. So vast is this input that the sea would soon be saturated with calcium bicarbonate were it not for two processes which release it from the sea water. First, in certain conditions of temperature, salinity, acidity, and pressure calcium carbonate can be precipitated directly onto the sea bottom, mostly in tiny particles. Second, far more significant, almost all marine organisms extract dissolved calcium carbonate from its bicarbonate solution for use in building shells, other protective coverings, or skeletons.

At this point secondary cycles tie into the principal one. Locked in an oyster shell or a coral-head the calcium carbonate may be liberated by another animal which devours

> "Rain and wind patiently gnaw away at the deposits of calcium carbonate on the land, washing them into the sea in solution at the rate of 500 million metric tons each year."

or bores into it. Or the shell may sink into colder, high-pressure areas of the oceans that are calcium-poor, there to redissolve.

(Practically no calcium carbonate is to be found deeper than 16,000 feet.)

The calcium carbonate story possesses ramifications reaching beyond the creation of coral reefs. Free bicarbonate ions comprise one of the ocean's strongest buffering agents, reacting with other minerals on the sea-bed for the maintenance of relatively stable ratios and acidities—during such "acid" events, for example, as the eruption of an undersea volcano. Indeed, it was carbonates and other carbon-containing substances in the sea which combined 3 or 4 billion years ago to form those precursors of life—the first organic molecules—that in time developed the capability of ultilizing sunlight for the synthesis of their own food, in more time evolved senses and other specialized organs, in still more time evolved us! Here yet another carbon cycle became established.

Man plays two roles in the carbon cycles. Like any other animal, he consumes organic (carbon-containing) substances and exhales CO_2. But in recent years he has supplemented this "natural" contribution by large-scale conversions of pure carbon into CO_2 through industrial operations that burn the fossil fuels, coal and oil. In the past 50 years the percentage of CO_2 in the atmosphere has risen measurably. Probably the ocean will save man from his peril—again! The amount of calcium carbonate precipitated to the ocean floor is a function of the amount of CO_2 in the air. And so, if all else fails, the oceans can probably ensure an acceptable range.

Coral reef. Hard and soft coral, various types of fishes and worms, crustaceans of all varieties, and sponges abound in this submarine scene.

Fishing the Seas

A / Sport fisherman. *Strictly for recreation—although the fisherman can make a meal of his catch. Still, sportsmen must not deceive themselves into believing that the sea is inexhaustible.*

B / Spear fisherman. *Also a sportsman, armed to kill, but giving his prey more of a "chance."*

C / Native fisherman. *Dependent on the sea for sustenance, he takes only what he needs.*

D / Commercial fisherman. *Unfortunately, often by sheer negligence, these fishermen will endanger whole species.*

E / Squids attracted to light. Commercial fishermen often use this method to attract fish.

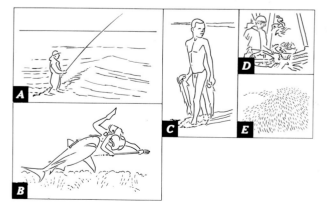

Chapter IX. Seasons and Provinces Of the Ocean

There are seasons in the sea, provinces as on land. And as on land, the sun's energy available at a given latitude together with the depth (or altitude) combine to produce the characteristic climates of the provinces. And in the oceans these are three-dimensional provinces, some of which interact with the land or the seabed, some of which are purely water provinces.

Today we barely have enough sounding profiles to outline a comprehensive geography of the sea. The continents are fringed

> "There are long, narrow abyssal trenches into which Mount Everest could be set and still have 6000 feet of water above its summit."

by littoral provinces, which extend underwater as the *continental shelf,* a vast province totalling 8 per cent of the surface of the oceans, ranging to the "drop-off-line" about 600 feet deep. A steep slope leads to the *continental margin,* then to the immense *abyssal plain,* 13,000 to 20,000-feet deep, dotted with volcanic islands and seamounts, streaked by mid-ocean ridges and rifts where the tremendous forces that push the continents apart are generated. Deeper are carved long, narrow abyssal trenches into which Mount Everest could be set and still have 6000 feet of water above its summit.

This general picture is made alive by rivers flowing from land, cutting canyons, pouring nutrients, pollutants, and sediments which trigger slumps and turbidity currents; by oceanic currents resulting from the density of the water and the rotation of the earth; by "vertical rivers" of cold water upwelling toward the surface; by constant intake of gases from the atmosphere and salts from the continents and the bottom; and by living creatures themselves, mainly plankton. Under the sea, the seasons can be as spectacular as on land.

The whole food chain responds to three dimensions as well as to a very high number of physical and chemical influences. Large commercial fishes like tuna, entire populations of sea birds, and marine mammals migrate in spectacular permanent exodus. For centuries these fantastic concentrations of sea life, achieved after millions of constructive years of unimpeded evolution, have enriched fishermen and gave birth to the myth of the inexhaustible mother sea. In fact, the average weight of organic matter to be found in a cubic mile of ocean water is to be counted in mere pounds, not tons.

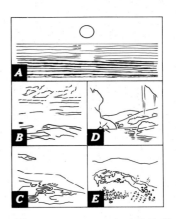

A / Sunset. *The serene sea at nightfall.*

B / Sea smoke. *Warm air over cold water.*

C / Fresh and salt water. *A river flowing to the sea.*

D / Spring. *When springtime comes to cold Arctic waters, marine life blooms.*

E / Bahamian trench. *Islands in the tropics standing besides a deep abyssal trench (dark blue) as seen from a satellite photograph.*

Ocean Geography

The Atlantic. Like a spine up the middle of the Atlantic Ocean, the Mid-Atlantic Ridge snakes its way from the southern pole of the world to its northern. Probably it represents, geologists think, the crunch-line of gigantic crustal plates underlying the ocean floor that collide with each other as volcanic forces from earth's mantle set them in slow motion. Along the edges of the continents appear the continental shelves. Between Ridge and shelves are the ocean's abyssal plains.

The Pacific. The gigantic floor of the Pacific Ocean is laced and interlaced with ridges and fracture lines running at 90 degrees to the ridges. Scattered over the ocean plain are mountainous islands, sea mounts (mountains beneath the surface), atolls (islands sinking at the same rate that coral reefs are growing on them, forming the typical circular lagoons), and guyots (flattened sea mounts, also beneath the surface)—all souvenirs of the tremendous volcanic activity in this region.

93

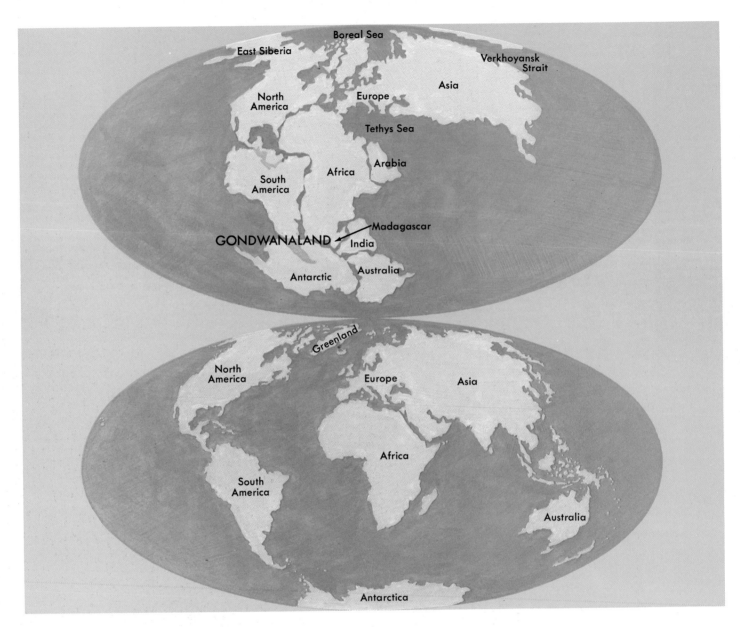

Continental Drift

In 1920 a German scientist, Alfred Wegener, published a theory he'd proposed for a number of years—that the land areas of earth had originally been one mass and that during the last 200 million years this supercontinent had broken apart and drifted to the degrees of longitude and latitude its parts now occupy. Today the theory of continental drift has gained acceptance. How do continents drift? To imagine such a thing, one must consider the solid land we inhabit as a light froth riding on earth's crust like bubbles riding in a pot of boiling liquid. In this pot the hot liquid moves to the surface and pushes the bubbles to the outer rim where cooler liquid descends. Essentially the same thing is happening to the earth. Heat from the core escapes along the central rift of the ocean and the upthrust material pushes the lighter continent apart.

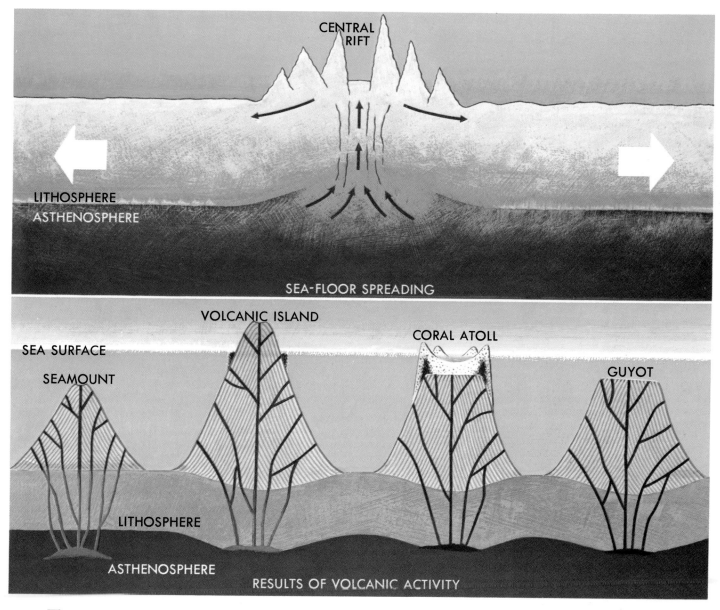

CENTRAL RIFT

LITHOSPHERE
ASTHENOSPHERE

SEA-FLOOR SPREADING

VOLCANIC ISLAND

CORAL ATOLL

SEA SURFACE

SEAMOUNT

GUYOT

LITHOSPHERE

ASTHENOSPHERE

RESULTS OF VOLCANIC ACTIVITY

Forces from Earth's Mantle

Islands of volcanic origin are found the world over. Among the most commonly referred to are the islands of the Pacific, especially the Hawaiian Island chain which extends for several thousands of miles. Some of these islands still readily display to the surface observer their volcanic origin, but others are more difficult to identify: for example, the coral atolls of the Western Pacific. But what is their origin? A popular theory today describes the earth's surface as made up of six large blocks, or crustal plates, similar to inverted saucers. Between these blocks we may find fracture zones which act as vents, permitting molten rock (basalt) to escape from the earth below the brittle crust. As this escape continues the submarine volcano grows upward, perhaps to a height sufficient to break through the surface, thus becoming an island.

Ocean Currents

The ocean's waters are constantly moving in a complex pattern of currents controlled by three forces: the wind, the rotation of the earth, and the differing densities of warm and cold water.

Winds blowing over the great expanse of open sea move the surface water in ripples, then in waves, and finally in swells. Currents move in the direction of the prevailing wind until the force of earth's rotation causes them to spiral off, to the right in the northern hemisphere, left in the southern.

Currents have an important role in regulating the world's climate by moving large masses of warm water to cold latitudes and cold water to warm ones. The climate in western Europe is far milder than that in similar latitudes because the Gulf Stream moves sun-warmed tropical water past its shores. Fog banks form off the coast of Southern California when warm surface waters are blown to sea.

Cold Arctic water sinks beneath the oncoming Gulf Stream and travels south in a countercurrent under it. This churning of the ocean's waters mixes them so effectively that the proportion to each other of dissolved salts is constant in all seas.

Speed of the currents varies, but it is estimated that the Gulf Stream moves at a maximum of six miles an hour as it sweeps past Florida and flows north. The cold countercurrent from the Arctic travels slowly along the bottom and may take several years to reach the Equator.

The Mediterranean Sea is almost landlocked, but it is still a part of the complex current system of the Atlantic. At the Straits of Gibraltar the Atlantic water flows east at a rate of about three miles an hour. In this arid region, sun and wind evaporate

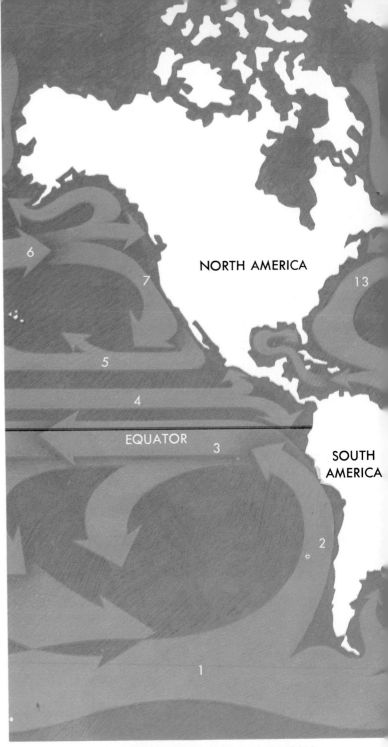

WARM CURRENTS

great quantities of surface water, leaving dense salty water that sinks to the bottom where it is pushed along to the west in a countercurrent.

At the center of the whirling water of the North Atlantic lies a quiet, calm sea called the Sargasso Sea. Floating on this sea is a large meadow of seaweed, the *Sargassum*.

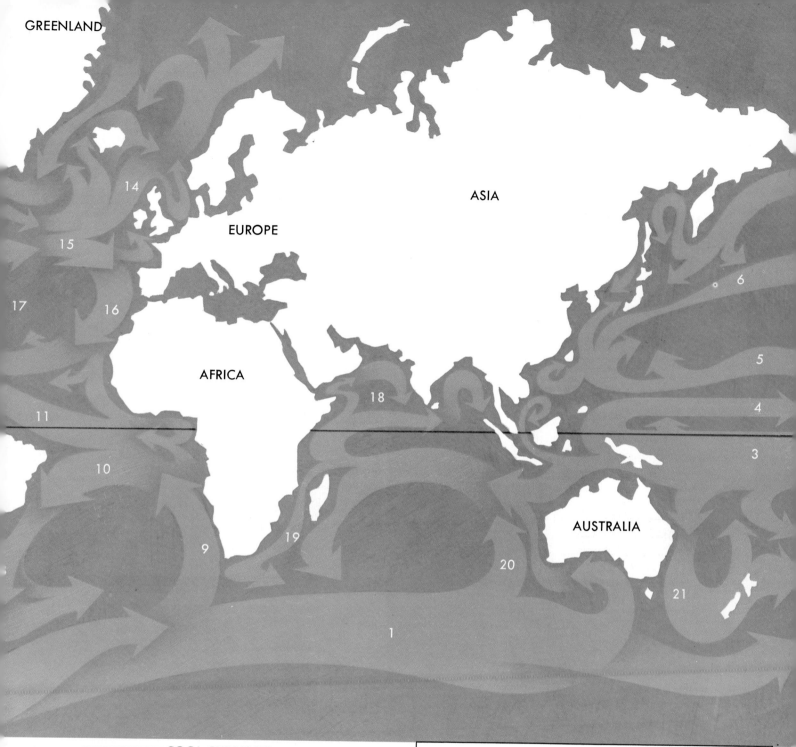

GREENLAND

ASIA

EUROPE

AFRICA

AUSTRALIA

14
15
17
16
11
10
9
19
20
21
1
6
5
4
3
18

COOL CURRENTS

Global circulation: *Warm currents circulate clockwise north of the equator and counterclockwise south of the equator. Cold ocean currents do the opposite. The results include a marked effect on the climates of land areas adjacent to oceans. For example, the Gulf Stream tempers the climate of the American East Coast. And the California Current keeps places like northern California, Oregon, and Washington not too hot in summer and not too cold in winter. The differing water temperatures also act as barriers to some fish.*

1 ANTARCTIC WEST WIND DRIFT	12. NORTH EQUATORIAL CURRENT
2. PERU CURRENT (HUMBOLDT)	13. GULF STREAM
3. SOUTH EQUATORIAL CURRENT	14. NORWEGIAN CURRENT
4. EQUATORIAL COUNTER CURRENT	15. NORTH ATLANTIC CURRENT
5. NORTH EQUATORIAL CURRENT	16. CANARIES CURRENT
6. KUROSHIO	17. SARGASSO SEA
7. CALIFORNIA CURRENT	18. MONSOON DRIFT
8. BRAZIL CURRENT	(SUMMER EAST, WINTER WEST)
9. BENGUELA CURRENT	19. MOZAMBIQUE CURRENT
10. SOUTH EQUATORIAL CURRENT	20. WEST AUSTRALIAN CURRENT
11. GUINEA CURRENT	21. EAST AUSTRALIAN CURRENT

Vertical Realms Of the Sea

Man swims through an ocean of his own —an ocean of air. He is as much a slave to it as any fish is to the ocean. I suppose most human beings are physically happiest around sea level. A healthy individual without cardiac difficulties can exist comfortably enough at the 8000-10,000 foot elevations of the Colorado mining camps. But at 17,000 feet, the altitude of some of the sky-high villages of the Andes, modifications have been made to the blood, the lung-capacities, and the hormonal system of the Indian inhabitants. The story is roughly the same in the ocean—with the reminder that water is 800 times denser than air and less transparent.

We look at the ocean depths in two ways: the "geography" of the earth underlying them and the "brightness" of the water spaces within them. First comes the tidal realm. Then come the continental shelf provinces — perhaps the most important area of the ocean, extending out to the drop of the continental slope where the true ocean begins. The shelf accounts for about 8 per cent of the earth's oceanic surface and goes out as far, approximately, as the depth of 600 feet. It is virtually another continent the size of Asia. At the margin of the con-

> "In the forlorn abyssal pit— 35,800 feet down—the crew of the *Trieste* found only a few odd animals an inch or two long. But they were still life!"

tinental shelf the continental slope steeply falls to the abyssal plains: immense undersea mountain-ranges and rifts, volcanoes sitting both above the surface (volcanic islands) or below it (seamounts), colossal trenches like those of the "ring of fire" around the Pacific Ocean many of which could easily swallow up Mount Everest and the Grand Canyon.

So far as light is concerned we divide the ocean into three zones: the sunlit zone, inhabited by plants and animals, which extends to about 600 feet or roughly the maximum depth of the continental shelf; the twilight zone, populated only by animals, from 600 feet to 4000 feet; and the black abyssal regions with their few and highly specialized animals. The bulk of life in the sea is concentrated in those waters that have some access to the sun: the sunlit surface waters inhabited by the countless constituents of the plankton, both plant and animal; the shallow seabed near the shores where stationary or crawling forms abound; the open sea just beneath the surface waters which is the domain of the oceanic fishes and mammals. And within the life of the ocean itself we also find the plankton layers and those mysterious deeper layers of small animals, rising and falling daily.

Perhaps the most interesting creatures of the sea are those which have at least partially liberated themselves from their "zone": the seal which can dive as far as 1000 feet; the dophin which can dive as deep and stay as long; the mighty sperm whale which can dive for one hour and reach the depth of nearly a mile. Beyond this depth life in the ocean thins out, and all is dark and silent. Almost all, I should say. When in 1960 Jacques Piccard and Don Walsh in the bathyscaph *Trieste* reached the bottom of the Mariana Trench, 35,800 feet deep, they found life even in that forlorn abyssal pit. Only a few odd animals, an inch or two long, subsisting on rare scraps of refuse which had drifted seven miles down—but still life!

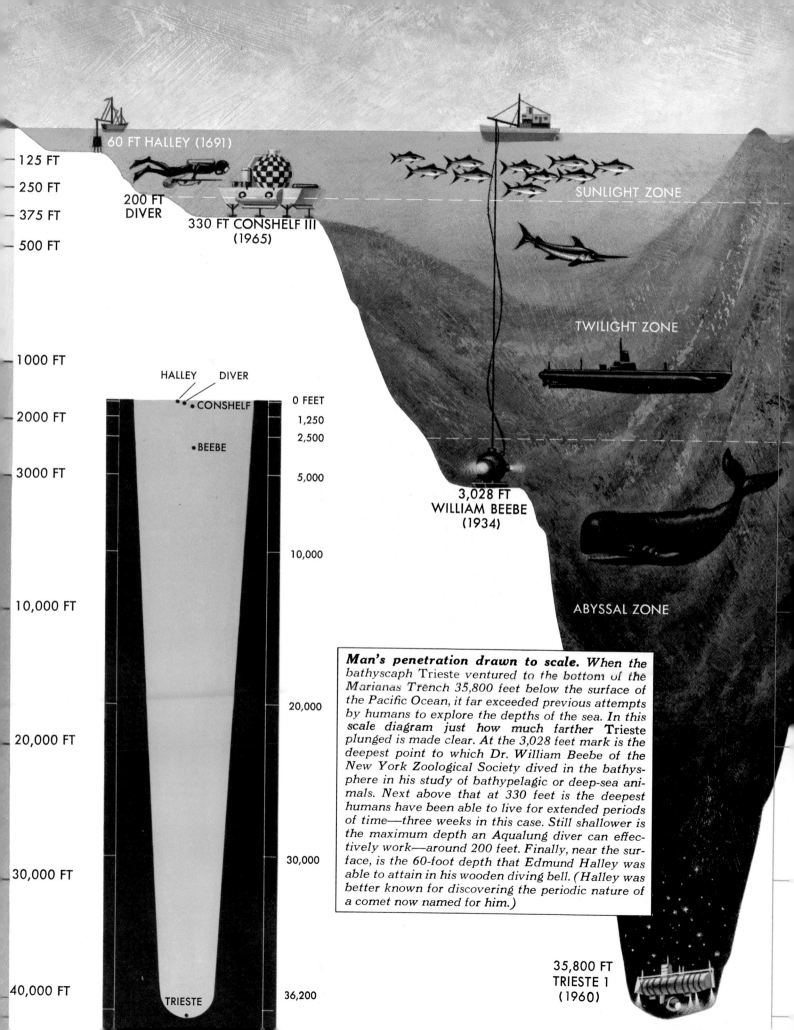

60 FT HALLEY (1691)

125 FT

250 FT

375 FT

500 FT

1000 FT

2000 FT

3000 FT

10,000 FT

20,000 FT

30,000 FT

40,000 FT

200 FT DIVER

330 FT CONSHELF III (1965)

SUNLIGHT ZONE

TWILIGHT ZONE

ABYSSAL ZONE

3,028 FT WILLIAM BEEBE (1934)

35,800 FT TRIESTE 1 (1960)

HALLEY   DIVER

CONSHELF

BEEBE

0 FEET

1,250

2,500

5,000

10,000

20,000

30,000

36,200

TRIESTE

Man's penetration drawn to scale. *When the bathyscaph* Trieste *ventured to the bottom of the Marianas Trench 35,800 feet below the surface of the Pacific Ocean, it far exceeded previous attempts by humans to explore the depths of the sea. In this scale diagram just how much farther* Trieste *plunged is made clear. At the 3,028 feet mark is the deepest point to which Dr. William Beebe of the New York Zoological Society dived in the bathysphere in his study of bathypelagic or deep-sea animals. Next above that at 330 feet is the deepest humans have been able to live for extended periods of time—three weeks in this case. Still shallower is the maximum depth an Aqualung diver can effectively work—around 200 feet. Finally, near the surface, is the 60-foot depth that Edmund Halley was able to attain in his wooden diving bell. (Halley was better known for discovering the periodic nature of a comet now named for him.)*

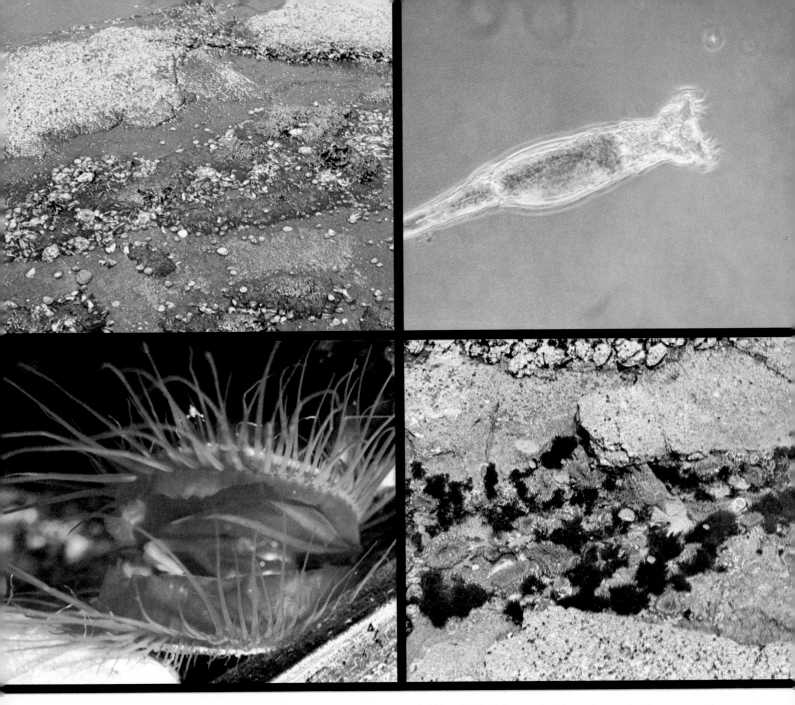

Life In the Tidepools

Along the edges of the sea, where rocky out-crops form ridges and hollows, life abounds in tidepools that form when tides rise and fall, when surf ebbs and flows. These pools harbor algae and animals, the former pro-viding food for the latter. And the latter feeding on each other sometimes. Mostly the animals are those that can evacuate the pools when they become dry on the outgoing tide—molluscs, fish, crustaceans.

A / Typical tidepool. After the water has run out, algae, seaweed, and molluscs are revealed.

B / File shell. These slow-moving scallops have tentacles which extend from their shells and pro-tect their eyes. The tentacles cannot be withdrawn completely.

C / Rotifer. Fast moving, microscopic, transparent creatures which occur in a variety of remarkable shapes. The largest of these glass-like creatures is about 1/50 of an inch long.

D / Sea-urchins. Brittle tentacled animals which move slowly. Some of the tropical species give off a

poison. They range from one-and-one-half to ten inches in diameter.

E / Starfish and mussels. *Common starfish have five to ten arms or more. Most of them have spines with tiny pincers. The mouth is underneath the disc part of the body. Starfish feed off shellfish and destroy tons of them yearly. Mussels are prized as food and are sometimes as large as six inches long. They can, however, be poisonous.*

F / Grunts. *These noise-making fish average about a foot in length. Some can change color for camouflage.*

G / The return. *Water returning to the tidepool.*

H / Nudibranch. *Meaning "exposed gills," these shell-less molluscs are predatory animals, feeding occasionally on sponges.*

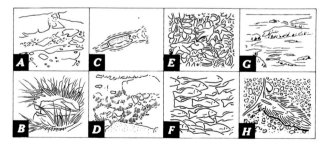

Life In the Coral Reefs

The reef community, in general, constitutes one of the oldest, most complex, and most tightly integrated of all ecosystems on this planet. Essentially, the reef is an offshore rock mass with a crest never much deeper than about 60 feet. Several types of reefs are found in the world's oceans: the fringing reef (in shallow water along coastlines), the barrier reef (running parallel to the shoreline), and the atoll (a circular island enclosing or partially enclosing a central lagoon).

A / Big-eye in coral overhang. Seaward of the back reef we come to the shallow reef flat where corals grow to a height equal to the level of the sea at low tide. The flaf—with holes, shallow depressions, and small tunnels—provides an uncountable number of hiding places for small fish and crabs, sea urchins, sponges, sea fans, and fragile corals.

B / Shoal of herring. A reef is constituted of two or more zones, adding up to an environment which can sustain many different types of marine life.

C / Anemone. The back reef of the fringing reef system is a shallow lagoon-like area carpeted with sediments from broken skeletal systems of long-dead reef occupants. But we find occasional outcropping of patch reefs formed by isolated colonies of coral which break up the otherwise smooth reef.

D / Sea star. *The famous "crown of thorns," which not only feeds directly on the coral animals but is increasing in such numbers as to constitute a major ecological disturbance throughout the Pacific Ocean.*

E / Angelfish and clown anemones. *As we swim seaward over a reef we come to a buttress zone: the outer limit of the reef system. Here the ocean floor generally slants upward at an angle of 45 degrees or greater to form a wall against the mighty sea.*

F / Grunts. *Many fishes live some of their lives in the open ocean but more in the warm shelter of a great reef.*

G / Anemone. *These hungry, but delicate, animals can survive only in the protection of the reef wall.*

H / Yellowtail jacks on reef. *Because of the force unleashed by oceanic waves the buttress zone is made up of the more durable corals, including tree-like structures whose branches form in line with the waves.*

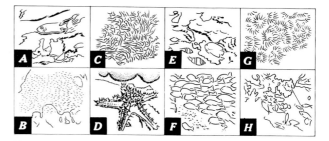

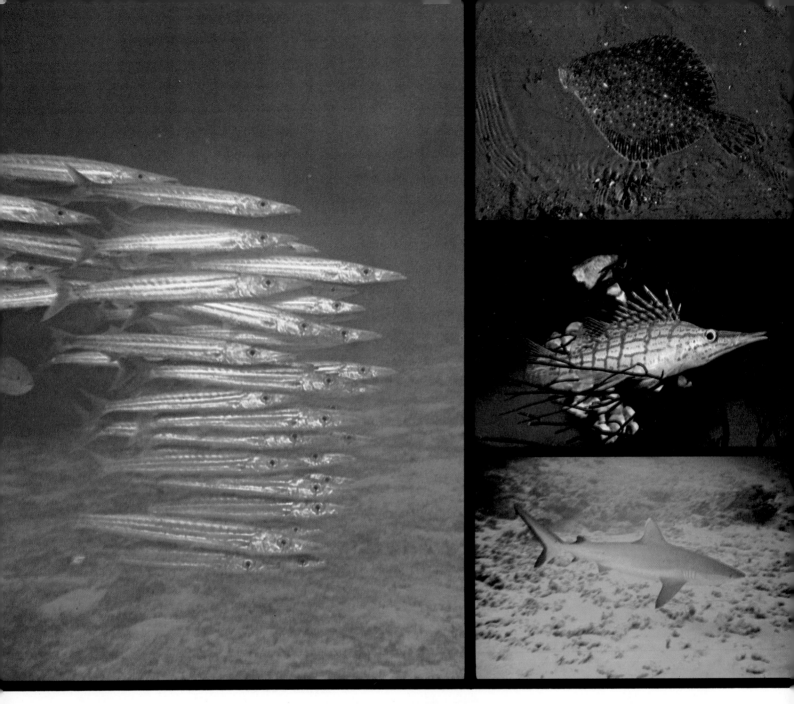

Life On the Continental Shelf

Along the edge of the world's continents is a shelf of land that extends into the sea to a depth of some 600 feet. On these continental shelves is concentrated a large portion of the plant and animal life of the sea. Here sun streams down and nourishes that which itself provides nourishment. And as algae flourish in these waters over the shelf areas, so does the animal life.

A / School of young barracuda. These large (five or six feet long) fish have been known to attack man. They feed by ripping with their teeth through a school of smaller fish and then returning to eat the pieces.

B / Sand dab. Both of this bottom fish's eyes are on its left side. It is not particularly large, rarely growing to longer than 15 inches, and weighing a maximum of two pounds. It has a large mouth which might indicate that it feeds on live prey which it surprises.

C / Birdfish. This agile, elusive fish lives in rocky reefs in the Indian and Pacific Oceans.

D / Reef shark. *This killer, sometime weighing in at 250 pounds, feeds at night. It is believed to slaughter netted fish, and to lay waste to whole schools of bluefish.*

E / School of mature fish. *Like Bermuda chubs, these fish school near reefs on the continental shelf.*

F / Sting ray. *This is a flat fish with a long tail housing a sharp, poison-filled spine which it can swing in any direction if it feels threatened.*

G / Diver in coral reef. *Here a diver descends to the serenity of a coral reef and disturbs the local inhabitants.*

H / Investigating. *Divers are here looking over the wreckage of a World War II Japanese plane, on which marine growth has begun to take root.*

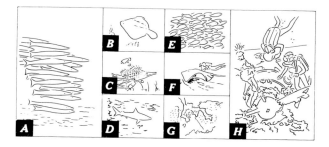

Life In the Open Sea

Open sea—land of contrasts, contrasts of abundance and scarcity. The clear blue transparency of open ocean waters is the color of liquid desert just as sable gold is the color of the Sahara Desert. As in the Sahara, the desert of the open sea is patched by oases of life—huge schools of fish, sprats, sardines, flying fish. These fishes gather in the widely separated fertile areas of the open sea that abound with phytoplankton, the drifting microscopic plants on which they or those they prey on feed.

A / Sun on the open sea. *Beneath the ocean's placid surface teem countless millions of sea creatures—some of which are described here.*

B / Swordfish. *These flat-billed fish swallow their victims whole. They will attack anything, often with no apparent purpose. Broken-off swords have even been found in the "skins" of whales and submarines!*

C / Diatom. *Here in the open sea, trillions and trillions of tons of plankton (both vegetable and animal) are produced and dispersed.*

D / Shark. *Cruising the open sea at an average of 20 miles per hour, this predator seeks prey.*

E / School of fur seals. *Migrating from all points in the North Pacific to the Pribilof Islands off Alaska for breeding, these mammals feed on fish and squid.*

F / Whale. *A barnacle-encrusted whale surfaces to exhale and catch a fresh breath of air. These huge mammals often travel in "gams."*

G / Barracuda and jacks. *The barracuda is perfectly streamlined for cruising open water in mostly leisurely fashion. The jacks are built for sustained high-speed swimming.*

H / Bottle-nosed dolphins. *These intelligent mammals often travel in small family groups, sometimes in "pods" of hundreds.*

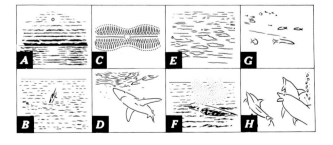

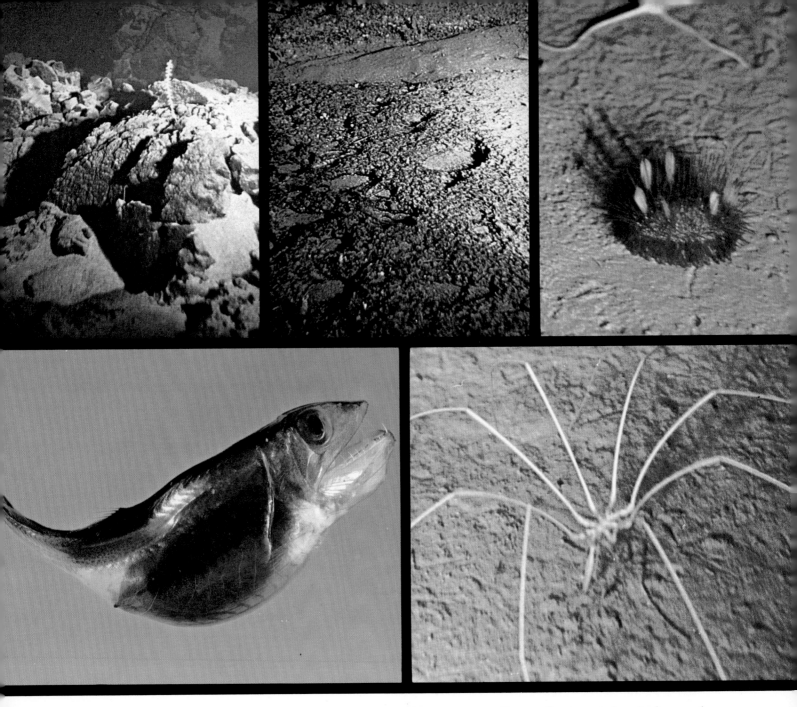

Life In the Deep Sea

In the deepest part of the ocean, known as the abyss, the water is ice-cold and the water pressure is greater than 1000 times atmospheric pressure. There is no light from the sun, but almost every animal there is bioluminescent, so that the abyss is alive with luminescent dots. It is not unlike the sky on a clear, dark night. The contours of the abyss are as varied as those of the continents, with steep mountains and mountain chains, valleys, and vast flat plains. Onto this primeval landscape of the ocean floor has fallen a steady rain of sediment. A study of these sediments reveals layer upon layer of different materials that have drifted downward onto the sea floor over hundreds of millions of years. Geologists often can translate this information into a history of our earth.

A / The "ring of fire." Here, on the ocean's bottom, volcanic lava helps to form the seascape.

B | The abyss. Resembling the lunar landscape, this region goes as far down as seven miles! It differs from the moon's surface in one very important way, however—life exists here.

C | Brittle-star and sea-urchin. Even in the deepest parts of the ocean, these relatives of sea-stars carry on daily life in a light-less environment.

D | Larval deep-sea perch. Its enormous teeth, even at this early stage in its development, mark this fish as a predator to be reckoned with.

E | Arrow crab. This crab delicately picks minute animals off the deep-sea floor as it stalks about on eight spidery legs.

F | Viperfish. The hinged jaw of the viperfish allows it to engulf large fish whole as they float lifeless to the bottom. In addition, the roof of its mouth contains 350 photophores which create a beacon of light with which to attract shrimp and little fish.

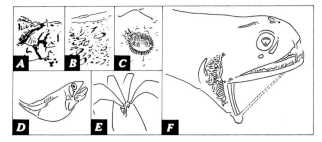

Birds Of the Sea

Within almost every group of animals are species which have returned to the sea. Birds are no exception. The sea is often a vast source of food. Various birds have adapted in a host of dimensions to life at sea. Webbed feet, elongated bills for burrowing in mud, and strong eyes are but a few ways birds have attuned themselves to their environment.

A / Adelie penguin. *One of only two species of penguin found regularly in Antarctica. They make long migrations through frigid waters for feeding and breeding.*

B / Horned puffin. *This uncommon bird has a remarkable triangular bill and is stubby and short-necked. It usually breeds from Maine to southern Greenland.*

C / Brandt's cormorant. *These large birds (body length 30-36 inches) sit upright and swim with their bills pointed upward. The Japanese keep cormorants to fish with!*

D / Flock of shore-birds. *Some birds at the edge*

of the sea find their food in the shallows and on the tidal flats.

E / Humboldt penguins. Living along the rocky cliffs of South America, these birds fish in a cold-water current that courses up the coast carrying with it many nutrients and fishes.

F / Guano birds. The source of a large Peruvian fertilizer industry, these birds feed on a fish called anchovetta which also provides Peruvian fishermen with their greatest resource.

G / Blue-faced booby. Found in the Caribbean, this two-foot-long bird makes spectacular plunges into the sea for its meals.

H / Double-crested cormorants. With its black body and orange throat, this is the only cormorant found on lakes and inland rivers.

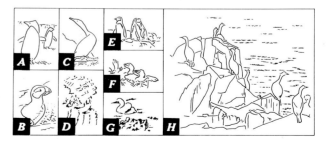

Mammals Of the Sea

The sea is a difficult place to live for any warmblooded, air-breathing animal. Legs are useless, water quickly absorbs body heat, air can only be obtained at the surface, and the dense liquid medium is extremely difficult to travel through. In spite of this the development of an aquatic life style occurred a number of times in the geologic past and common features appear in differing groups.

A / Seals. *These animals are as happy on ground as in the water. Their short, thick fur has all-too-often been used to clothe people. But as we can see from these photographs it looks good on them too, so really should be left where it belongs.*

B / At home. *The seals are seen here on Seal Island in South Africa.*

C / From the **Calypso.** *After we had the seals on board with us for a few weeks, we could easily go diving and swimming with them.*

D / Elephant seal. *This baby seal is not waiting to feed. It is emitting a loud scream, which it does for*

attention and when angry. It weighs around 75 pounds at birth.

E / The battle. These two adult male elephant seals are fighting and screaming furiously in a battle over mating or feeding—probably the former. They have been known to weigh as much as 7000 pounds!

F / Manatee mother and baby. These vegetarians are docile, plain, and stupid. Their ancestors millions of years ago were elephants!

G / Bottle-nosed dolphin. Here a mother dolphin nuzzles her baby. These mammals can stay submerged for about 15 minutes.

H / Manatee and meal. In Florida manatees are protected and encouraged to clear the water-hyacinth-choked channels. It is one of the manatee's favorite meals as well as its calling.

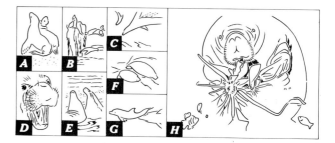

Chapter X. The Pharaohs Of the Sea

The Great Pyramid at Giza, built by the Egyptian Pharaoh Cheops between 2613 and 2494 B.C., one of the largest structures on Earth, stands over 400 feet high and contains about 2,300,000 blocks of stone, each weighing over two and a half tons.

It is a pygmy structure compared to those erected by the "pharaohs of the sea," builders of reefs and atolls, principal among which in our day is the coral polyp. A coral reef is made of the limestone skeletons secreted by innumerable coral polyps and left behind after they die, one on top of another. In 1837 Charles Darwin visualized a coral reef as developing around the top of a volcanic mountain that was subsiding into the sea at just the right rate to keep the coral polyps housebuilding at their favorite level —the top 150 feet of the tropical sea. After World War II, engineers from the United States Navy tested the theory by driving a shaft down through layer after layer of ever-more-ancient coral at Eniwetok Atoll. And at 4222 feet the drill hit into volcanic rock (proving the theory).

Corals are the chief animal reef-builders today, but many other creatures are associ-

> "Polyps of colonial corals are all interconnected by lateral attachments—so the living coral colony lies entirely above its skeleton, completely covering it."

ated in reef-communities. In the past, other species, now extinct, played the lead role; they are responsible for the older sections of reefs. The reef association of plants and animals in the tropical waters of the world is the most complex of all ocean ecosystems. It is also the oldest ecosystem in the world, more than 2 billion years old.

Nine tenths of a typical reef consists of fine sandy detritus, cemented by animals and plants such as the enveloping plates of calcareous algae. Physical and biological processes convert this stabilized detritus into limestone. Remains of dead reef organisms make a contribution to the detritus. This major component of the reef has a fabric different from the upward growing lattice of stony algal deposits and intertwined coral skeletons that forms the reef core.

At the top of the modern "reef community" pyramid, the stony coral produces a calcium skeleton. Large colonies of polyps feed at night—waving minute food into the pharynx. The skeleton is composed of calcium carbonate crystals—secretions producing a skeletal cup called a theca within which the polyp is immovably fixed. Polyps of colonial corals are all interconnected by lateral attachments. So the living coral colony lies entirely above the skeleton, completely covering it. Polyps live in symbiosis with microscopic one-celled plants (zooxanthellae) embedded in the animals' tissue, where they are nourished by nitrogenous animal wastes and where they add oxygen to the surrounding water. These promote the calcium metabolism of corals.

The complex reef community consists of other members—limestone-secreting algae, limestone-secreting families of sponges, monocellular foraminifera, microscopic colonial animals, spiny sea urchins, echinoderms, bivalves, clams and oysters, all of whose accumulated skeletons and shells contribute to the reef limestones.

Coral atolls. As seen from this satellite photograph, these coral rings are surrounding volcanoes which once were above the surface of the water but have since collapsed and been submerged.

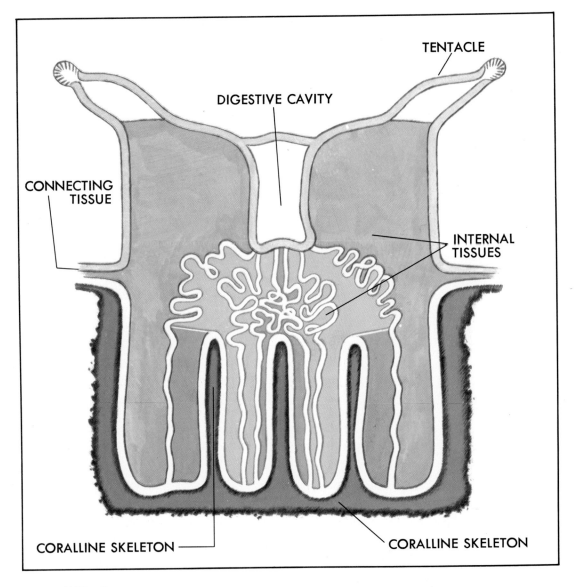

TENTACLE

DIGESTIVE CAVITY

CONNECTING TISSUE

INTERNAL TISSUES

CORALLINE SKELETON

CORALLINE SKELETON

The Coral Polyp

Most individual coral polyps are extremely small, but they are a colonial species and their colonies can attain considerable size. The animal can be thought of as a kind of self-service stomach. With its pharynx open to the sea, the coral waits for currents to wash something in the way of an edible object toward it. This the coral can receive passively or, using its tentacles, actively. In either case the food-particle is passed down into the central cavity of the animal, where it is ingested. What makes this humble little animal a "pharaoh" is that the lower part of the coral polyp's outer layer secretes a calcium carbonate skeleton, a cup called a theca within which the animal is immovably fixed. The polyps of the corals in a colony are all laterally interconnected—so that in a sense the colony itself, lying entirely above the skeleton and completely covering it, can be considered a single creature. The colony grows by means of the budding of new polyps. Countless numbers of these living, growing colonies have created the mountainous, infinitely variegated reefs we find everywhere in tropical seas.

A Stupendous Disaster?

Certainly two and maybe four times in the past 500-600 million years the reef community has "collapsed," with dozens of species of animals becoming extinct. After each of these collapses the reef community around the world has managed to reestablish itself with new balances, new species of animals in the typical "reef" association. Perhaps the most spectacular of these collapses was the fourth and, so far as we know, the last. This occurred about 65 million years ago at the close of the Cretaceous period. This was a period of great extinctions. Nearly a third of the families of all animals known in late Cretaceous times were no longer alive at the beginning of the next era, the Cenozoic. The reef community was not exempt. Two-thirds of the known genera of corals died out at this time, in addition to many other major groups. This late Cretaceous disaster also affected the land—this was the time of the end of the dinosaurs. Of the 115-odd genera of dinosaurs found in late Cretaceous fossil deposits none survived the end of the period.

To understand the reasons for such a stupendous disaster we must realize that the Earth has much changed its shape in these millions of years. Today's land area occupies about 30 per cent of Earth's surface. In the late Cretaceous almost two-thirds of today's land area was submerged under shallow oceans. The climate was warmer everywhere; there were no polar ice-caps; the weather of the whole world was of a piece. Then, at the end of the Cretaceous period, as a result of movements of Earth's crust, deep sea-basins were created which drained off the shallow oceans from much of the land. When the Antarctic ice-cap was formed some 20 million years ago, tremendous changes occurred in the climate, the weather, the currents of the oceans. What members of the reef community, including corals, were able to survive the long environmental change retreated to their present range: the tropical seas.

It is perfectly possible that there could be a fifth collapse of the reef communities. It would be accompanied by drastic changes in shorelines and weather. Man, with his technological ingenuity, could undoubtedly survive any geological vicissitude as a species. The only threats that could seriously endanger him are those that he would create himself. What happens today to the coral reefs all around the world is difficult to understand and it is too early to name it a new "collapse." But it is a very serious regression. The exuberant varieties of coral are disappearing from the northern and southern part of the Red Sea; from the Seychelles Islands to the Mozambique straits the fringing reefs are choked with sand and die. For several years most of the reefs in the Pacific Ocean have been badly damaged by a large coral-eating starfish, the "crown of thorns."

One thing is certain. If the coral reefs should disappear, the world would lose in variety and stability, and our grandchildren would

> "Certainly two and maybe four times in the past 500-600 million years the reef community has collapsed—with dozens of species of animals becoming extinct."

be deprived of the most beautiful sceneries the planet has ever produced. Who does not wish to swim or dive in the lagoons of the Caribbean or the Pacific—with gigantic brain corals standing in their gardens of sand, perhaps festooned with the mustard-hued "fire coral" on which every novice burns himself at least once.

Reef Builders

Found in tropical and sub-tropical waters, these stone-like animals are not rocks, as it would appear, but tremendous colonies of tiny marine animals in limestone castles. They are sensitive to light and need phytoplankton to feed on. Their protective "shell" remains long after the polyp dies and over centuries the colonies build up, becoming coral reefs. Sea fans and sea whips, however,

118

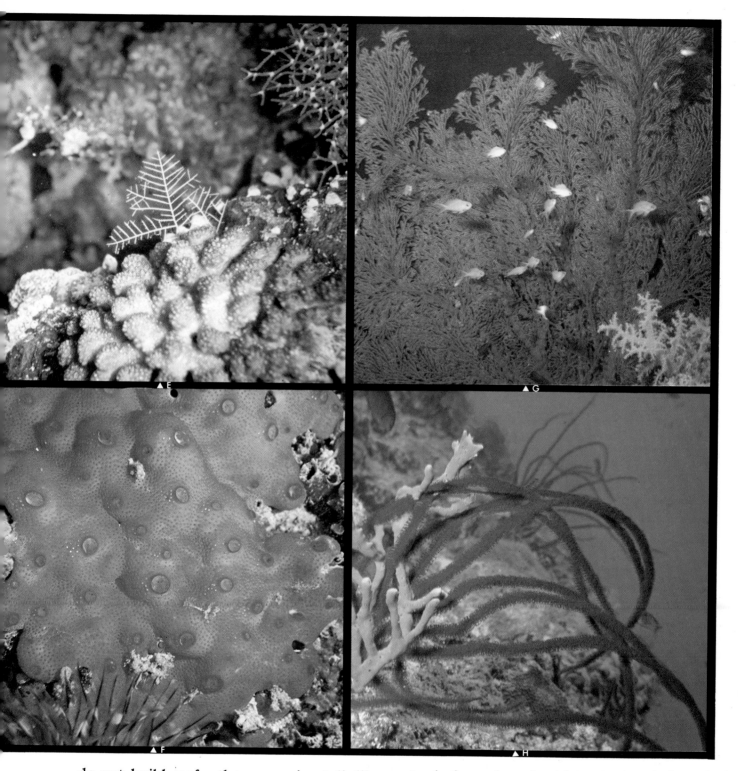

do not build reefs; they grow in stalk-like colonies into shapes of fans and trees. These are seen in boxes G and H on page 119. Box E on page 119 shows a hydrozoan in the form of a Christmas tree amid a different type of coral. Box A on page 118 shows a typical coral reef with many varieties of coral and a school of fish swimming serenely by. Box D is a brain coral, so named because of its resemblance to the human brain. Once the polyp dies, this coral leaves a white skeleton.

Tales From the Maldives

It takes centuries for a coral reef to grow. The work is occasionally undone within a fraction of that time by man or nature. An example is found in the geographic history of the Maldive Islands, a group of atolls in the Indian Ocean west of the tip of the Indian subcontinent. Those islands, surrounded by coral reefs, were formerly a

> **"Within a 24-hour period one starfish can graze over a reef area twice the size of its six- to 12-inch central disc."**

bustling stop on the great spice routes of Europe's merchant princes. Today they are almost forgotten. Traditionally the islanders built their homes of coconut thatch. As the modern world began to catch up with the islands, large buildings and roads had to be built, and the only construction material available was that from the coral reef. Now the Maldivians harvest coral the year round, using large chunks of it as building blocks or breaking it up to make caulk. By destroying the fringing coral reefs, the Maldive people are unknowingly dooming their islands, for without a protective ring to shield the atolls from the pounding of trade-wind waves, the islands must themselves eventually disappear.

Thousands of miles to the east in the Pacific Ocean, Guam and other islands are similarly threatened. But this time man—the apprentice of the sorcerer—had to give an explanation. There had to be a scapegoat, and the animal indicated was the prickly starfish known as the crown-of-thorns. Even Australia's 1,250-mile-long Great Barrier Reef, the largest coral reef in the world, had been attacked by these creatures. Though it always was a member of the coral reef community, this predator of polyps was going through a mysterious population explosion. Today it has invaded the reefs of the Pacific, and one can easily spot its path, patches of bleached coral skeletons stripped of their tiny polyps. The dead coral is rapidly covered with a thick fuzz of algae that prevents new coral growth. Bereft of their living polyps, the reefs can easily be destroyed by the erosive action of the pounding sea.

But this is an oversimplified picture: within a 24-hour period one starfish can graze over a reef area twice the size of its 6-to-12 inch central disk. Of course, the solution is not for man to add one destruction on top of another and to attempt to eradicate the starfish population! Incidentally, if speared or cut in half, these crowns-of-thorns are capable of surviving and growing new arms. Islanders naively try to poison them individually or to collect them and bury them on land! Some scientists recommend importing their main natural enemy—a giant sea snail. One theory is that the population explosion is simply a natural phenomenon. In Australia it is believed that excessive collecting of the triton sea snail has permitted the starfish to reproduce abnormally. In Micronesia it is suggested that in the process of dredging channels, or dynamiting for fish, man has killed the coral or harmed it enough to alter the balance of the environment. Under normal conditions, most of the millions of eggs spawned by the crown-of-thorns are eaten by the live, healthy polyps.

Loss of the Maldive Islands. The coral reef protects the islands from strong winds and waves. In these photographs we see men unwittingly destroying their homes. They are taking huge pieces of coral from the water for use as building blocks.

Chapter XI. Riches From the Sea

Energy is obviously earth's crucial resource. All the energy on earth springs from a single mother-lode, the sun: a gigantic thermo-nuclear furnace in which the fusion of lighter atoms into heavier ones releases energy which radiates across a gulf of 96 million miles, bathing the earth and all that live on it in light and heat.

The sun has been shining on the earth since the planet was born—some 5 or so billions of years ago. And for perhaps half that period the sun's rays have fed into "energy accumulators" on earth's surface: life-forms, plant and animal, which store up solar energies in their leaves and tissues. So for 3 or 4 billion years the sun has presided over countless generations of living things whose dead re-

> "Sooner than we think we will have to reach out directly for the sun's energies, no doubt by the installation in space of orbiting stations to catch and concentrate its rays."

mains have sunk into earth's crust—there, after many millions of years, to be converted into our "fossil fuels," coal and oil. From the giant rainforests of the Carboniferous Period, from the corpses of uncountable trillions of animals large and small, come all the coal and oil on earth, with much of the energy the sun poured into them still locked in, awaiting the key of fire. At last man entered, digging holes in the ground, uncovering these caches of energy. He reared his new material-istic civilization on them. In 1900, 96 per cent of the world's industrial energy came from coal. Recently oil has challenged coal's dominance. But still today mankind's energy needs are met mostly by the fossil-fuels.

In our time the fossil-fuel resources of the ocean have been called upon to supplement the fast-fading stocks on land—with large-scale drilling enterprises on the continental shelves and sea-beds. But like all bank-accounts the fossil-fuel energy bank-account has its limits. With his population booming from 1.5 billion in 1900 to 3.5 billion in 1970 to a probable 7 billion in the year 2000, a population which furthermore is increasing its per capita energy demands at an accelerating rate, man can already fore-see only too clearly the day when he has exhausted earth's fossil-fuel legacy terres-trial *and* oceanic. Fossil fuels, quite rapidly, must grow rarer and more costly to get at.

The sea-bottom also begins to be mined for various minerals—phosphates, heavy met-als, complex manganese nodules. Valuable metal-salts, used in the manufacture of drugs, are extracted from sea-water itself. And a wealth of new drugs, such as anti-biotics, can be isolated and developed in the ocean. Man must accomplish these extrac-tions of riches from the sea with more cau-tion and finesse than he is accustomed to. You cannot simply rape the ocean—at least, not indefinitely. Every mine, every factory, every processing-plant, every new construc-tion of any kind destroys to a degree the natural environment around it. The oceans' greatest food resource is of course fish, of which about 60 million tons are landed a year. There is no question but that the sea's yield of fish can be multiplied many times if blind fishing methods are progressively re-placed by scientific "farming."

The sea. "In our time the fossil-fuel resources of the ocean have been called upon to supplement fast-fading stocks on land."

Exploitation Of the Sea

A | Oil well. *In the near future almost 50 per cent of the world's oil will come from the continental shelf.*

B | Food fish. *Sophisticated fishing activities in Arctic waters will soon threaten a number of marine species.*

C | Manganese nodules. *This valuable mineral is about to be taken from the seabed for use in industry.*

D | Underwater oil well. *Oil companies presently employ more divers than any other marine industry.*

E | Kelp. *Kelp is necessary to a healthy marine environment but also useful to man in the manufacture of paint and other commodities. Kelp harvesters and ecologists are working together to reforest kelpbeds along the California coast.*

F | Sea-weed for food. *In Japan, particularly, sea-weed is a popular addition to the everyday diet.*

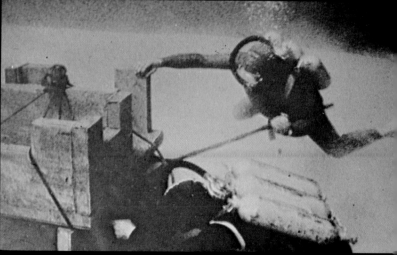

G | Fish farming. Here concrete structures provide growing surfaces for algae and corals which attract fish populations for farming by men.

H | Scallop. Important as a food animal, the scallop is also valued for its attractive shells which are sold for use as plates or ash trays.

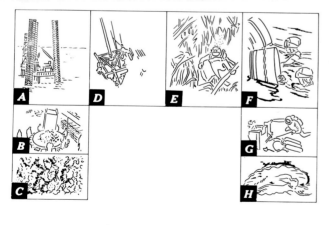

Chapter XII. Return To the Sea

In September 1965 a momentous phone-call in the story of explorations occurred between two young men at the bottom of the ocean—6000 miles apart. From Conshelf III, 328 feet down in the Mediterranean off the coast of France, oceanaut Philippe Cousteau talked to aquanaut Scott Carpenter in Sealab II, the United States Navy's undersea experiment-station 208 feet down in the Pacific off the coast of Southern California.

Conshelf III was a project designed to show that men can do more than simply manage to exist on the sea bottom. They can do work there, complicated work, more effectively than machines can do it. It followed up two earlier projects of ours. In Conshelf I, the first manned undersea station in the world, two divers, Albert Falco and Claude Wesly, submerged for a week in 35 feet of depth, breathing air. They became the first "oceanauts." Conshelf II, on the other hand, was a sort of undersea colony in the Red Sea, with a number of separate installations where five men stayed a month at 36 feet; two

> "In engineering Conshelf III
> I wanted to eliminate as far as
> possible the dependence on the
> surface of the oceanauts. I did
> not want their project, their lives,
> jeopardized by our failures."

lived a week in a heliox environment at 90 feet, worked in water at 165, and visited depths at 330 feet. With Conshelf III—a more ambitious operation involving 150 technicians and two ships—we aimed at greater depths and more specific tasks. We were determined that once they arrived on the bottom the six young oceanauts would demonstrate conclusively that the undersea environment is one to which man can productively adapt himself.

In engineering Conshelf III I wanted to eliminate as far as possible the dependence on the surface of the oceanauts. I did not want their project, their lives, jeopardized by our failures. Therefore, we supplied the Conshelf III "home" with every form of self-sufficiency we could think of. Conshelf III was spherical and capable of resisting inside or outside pressures of 20 atmospheres. Compression and decompression of the oceanauts were performed safely in the harbor. Then, the structure could be towed to the chosen site and sunk. Conshelf III had enough breathing gas and CO_2 scrubber—as well as succulent frozen meals prepared by Air France chefs—for one month. The only "umbilical cord" to the surface was a high voltage power cable. Because ordinary compressed air could not be used at this depth, they breathed "heliox," a mixture of oxygen and helium. Helium is a merry and mischievous element which carries along with it a bag of tricks—some amusing, some infuriating. More seriously, helium conducts heat more efficiently than air, so all objects on board cooled faster than was desired. The oceanauts had to wear insulating vests of a special material. They could not smoke —not because of any danger of fire but because cigarettes will not stay lit in a helium atmosphere. On top of everything else, although humans function well enough in helium, we found it almost impossible to keep the stuff from infiltrating our electronic apparatus, which it tended to drive crazy.

But as my son Philippe noted, "As soon as I venture away from the undersea house I am struck by one fact—we have lost the surface. It is far above us, out of sight, buried in treacherous night. The surface means death."

A Job of Ocean Work

Below you see a picture of Conshelf II, shrouded in the darkness of the sea at a depth of 36 feet. Sited at a much greater depth, Conshelf III was almost unphotographable. But it was a great success. We had our setbacks—you expect them in an experimental operation of this magnitude in the sea. On the fifth day after the "home" touched down on the sea bottom a storm rose which came near to tearing away what support systems we needed. The well-head for the oil-drill came bouncing down like a yo-yo. Various articles of equipment failed under the unprecedented stresses. As the days wore on the oceanauts themselves grew cold and fatigued. But they did their jobs magnificently.

We had proved our point that man can occupy and exploit the sea bottom—by developing new technologies of high-pressure breathing apparatus, by using our new incompressible insulating vests, by keeping to a minimum the life-support ties to the surface. Perhaps most significant of all, we had begun to breed a new sensibility, a new sense of confidence.

Conshelf II gleaming through the murk: "a sort of undersea colony in the Red Sea."

"Sea Lab to Conshelf: Are You There?"

In 1965 the first inter-undersea telephone conversation took place. Astronaut-Aquanaut Scott Carpenter from the United States Navy's Sealab II off La Jolla, California, spoke to French oceanaut Philippe Cousteau in Conshelf III off Cap Ferrat in the Mediterranean—6000 miles apart! The greetings exchanged between the two pioneering underwater stations were confused by static, language differences, and helium, the last of which causes the pitch of the voice to rise and become almost unintelligible. Now that man can live and work in the depths for weeks at a time, "saturation divers" have opened up millions of square miles of the continental shelf to man.

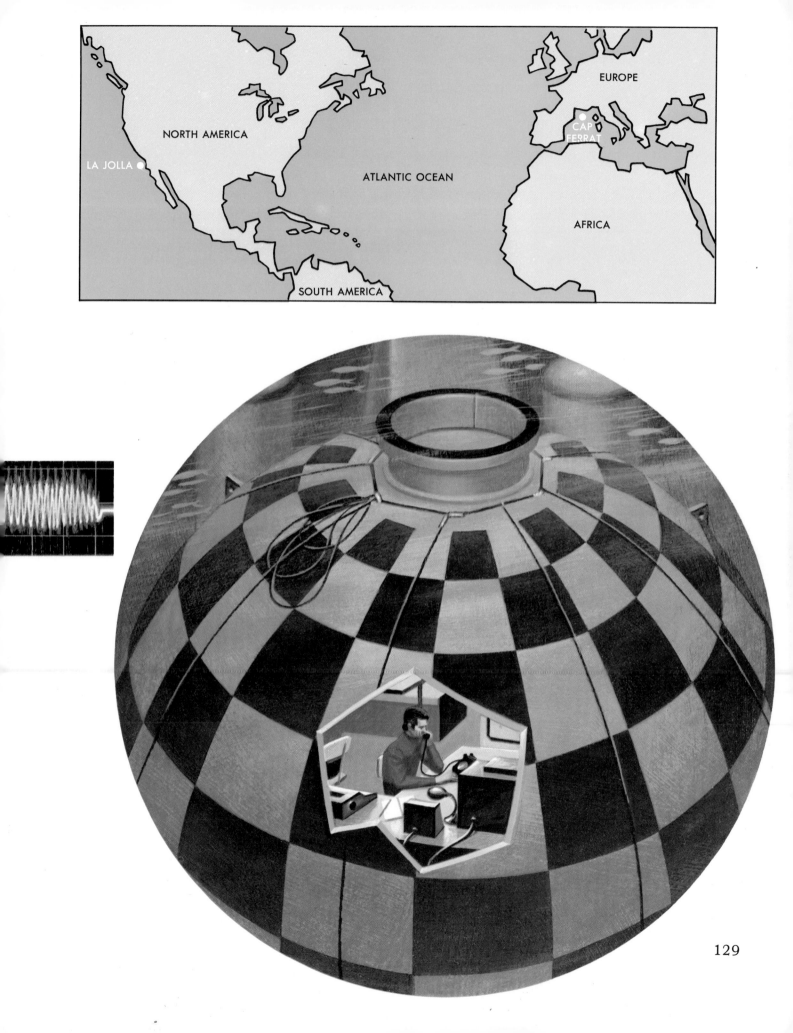

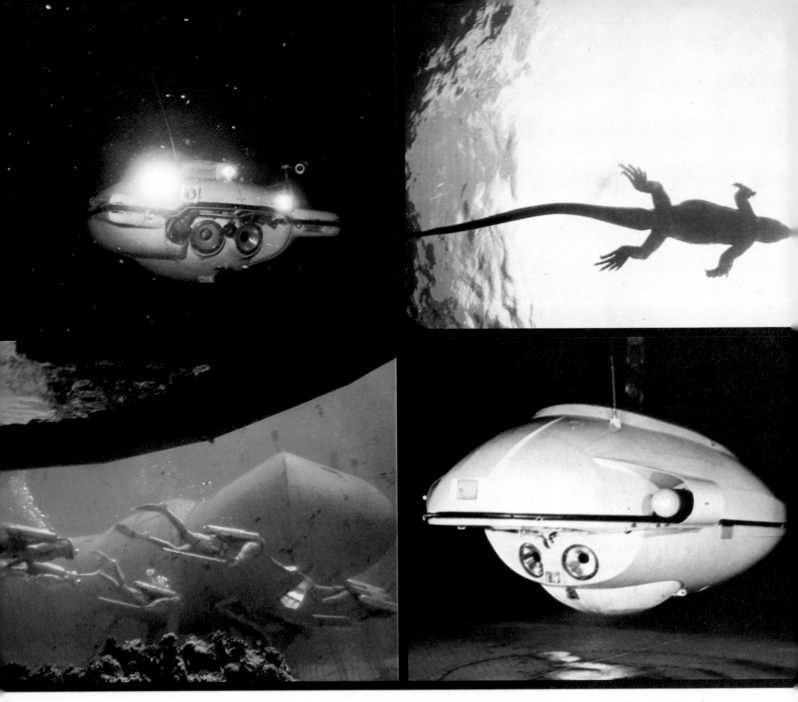

Invaders Of the Sea

From the days of the ancient Sumerian hero, Gilgamesh, who dived for seaweed that would give him eternal life, man has dreamed of returning to his original home in the oceans. Other animals have gone ahead of him: those familiar to us as well as some not so widely known. Some air-breathing spiders build houses under water in which they raise their young away from terrestrial predators. Man's return has not been easy. Although we have once touched the deepest part of the ocean floor, we have made "homes" only on the shallowest half of the continental shelf.

A / Diving Saucer SP 350. Built in 1959, this is one of the most successful and practical observation submarines ever built. It can take two men down to 1000 feet.

B / Starfish house. Five men lived, worked, and observed for one month in this submarine colony on the ocean floor.

C / Iguana. The Galapagos iguana spends most of its time on rocky islands, resting in the sunshine. It returns to the sea to graze on algae beds and is capable of remaining underwater for long periods.

D / Diving Saucer SP 3000. *Not yet in its working stages, this submarine will eventually allow men to penetrate and observe almost 10,000 feet down!*

E / Trieste. *This pictures the triumphant return of Jacques Piccard and Don Walsh after they had breathed air at atmospheric pressure while almost seven miles under the surface.*

F / The diving spider. *This air-breather was the inspiration for Sir Edmund Halley who built a diving bell in 1691. The spider brings air bubbles down from the surface on the hairs of its legs to fill the dome-like nest it has built underwater.*

G / Diving Saucer SP 1000. *This is a one-man submarine which is useful down to 300 feet.*

H / Interior of Conshelf II. *Life goes on here as it would on the surface but for the view through the picture window.*

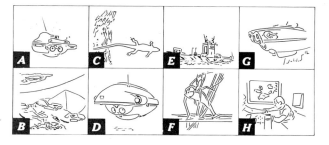

131

The Future

"Is it possible that the future of undersea exploration belongs to the wealthy governments of earth? I would hate to admit it. Up to this point my colleagues of countless undersea adventures, and for 100 years men like them from dozens of nations, have done the job. We have proved that man can accommodate himself psychologically and physically to the depths for long periods. We have proved that man can do both scientific and industrial work in the depths more efficiently than can machines.

"We see clearly what our next step is: to increase our 'staying power' on the sea bottom. Unfortunately, the means of taking this step involves equipment which is so costly—with special metals and electronic apparatus and elaborate life-support and nuclear-propulsion systems—that perhaps only rich bureaucracies can afford them.

"At the top of the next page you see the artist's rendering of the submersible Argy-

> "We see clearly what our next step is: to increase our staying power on the sea bottom."

ronete, at this writing under construction in Marseilles. In effect Argyronete will be a mobile Conshelf III. We call her an 'intervention submarine.' She differs from conventional submarines in having two pressure compartments: one, standard in submarines, remains at sea-level pressure, and houses the captain and a crew of five. The second chamber will house four diver-oceanauts and can be pressurized to match a depth as great as 2000 feet. The divers, their blood 'saturated' in this atmosphere, move easily in and out of a bottom opening hatch through the hull of the 'lock-out'

chamber, with no more umbilical connection to the submarine than the hoses through which their breathing mixture and communication cables pass. The advantages of Argyronete's mobility are obvious.

"Argyronete is progress. But perhaps we won't rest until we experiment with the gentleman you see swimming toward you from the bottom of the opposite page. Meet Homo Aquaticus. He is a human being who has been medically engineered into the adaptations took millions of years to accomplish with the dolphin. He has undergone a surgical operation which has replaced his lungs with a unit containing a special fluid that furnishes oxygen to his circulatory system. As a result, he cannot breathe a gas, he does not have to carry a clumsy air supply on his back or retain an air-hose link with a ship, he does not have to worry about the 'bends' or other symptoms of nitrogen narcosis or decompression trauma. He can travel as deep in the sea and stay as long as he likes, and make a quick return, with no ill effects. With the exception of periodic visits back to Argyronete, or a similar base, for the replacement of his oxygenated fluid unit and CO_2 absorbent, Homo Aquaticus is completely at liberty in the sea: to play, to dream, to farm whales and herd fish, to repair undersea machinery, to supervise research."

A / Argyronete. The submarine of the future will allow a diver to leave the submarine to work in great depths and return to it. Remaining under pressure, he can then be transported to another underwater site or back to shore.

B / Homo Aquaticus. In the future a diver may no longer require a breathable atmosphere. An operation could make it possible to bypass the lungs and oxygenate his blood directly.

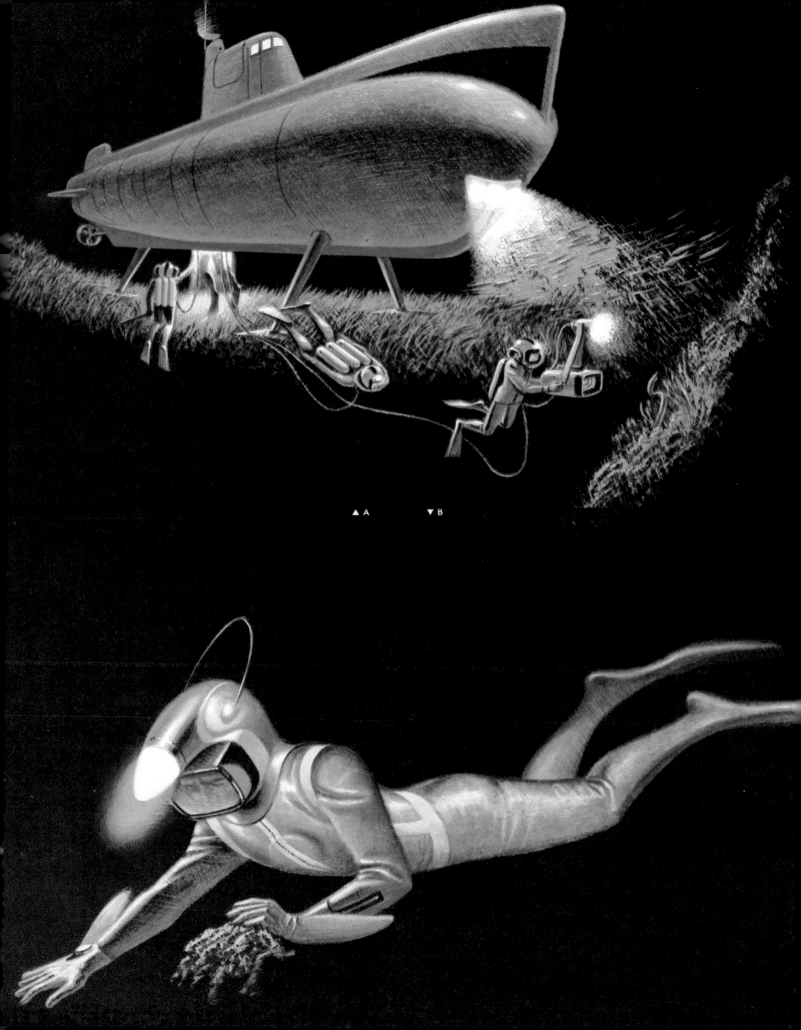

▲ A   ▼ B

Chapter XIII. The Sea in Danger

The human brain is capable of intellectual miracles. The human spirit is capable of moral miracles. When we listen to the Brandenburg concertos of Bach, when we think about Shakespeare's plays, Einstein's cosmologies, the chess strategies of Bobby Fischer, the historical unravellings of a

> **"The plain sensible fisherman knows that if you take fish out of a pond beyond the animal's capacity to reproduce itself you kill the pond for the next season."**

Michelet or a Marc Bloch—when we imagine the superb nervous systems behind these achievements: disciplined, indomitable, grappling with the inner relationships between antagonistic elements—we see what our species can aspire to.

But some part of man's capacity for common sense and for long-range planning seems to desert him when he becomes just a member of a group like a "collective," a "society," a "nation." In groups men and women seem to lose pieces of their individual foresight and humanity. In groups we are more easily led around by the noses, whipped up into artificial needs, emotional frenzies, or political panics of one description or other.

Paradoxically, in groups we may at the same time become self-centered, indifferent to the future, stupid. The plain sensible fisherman knows that if you take fish out of a pond beyond the animal's capacity to reproduce itself you kill the pond for the next season. All farmers understand the same thing about their lands. But somehow man as "urban consumer" fails to realize that the

modern fishing industry is doing precisely this in earth's central oceans. The plain ordinary man knows that shooting polar-bears from helicopters, for example, cannot be accounted a sport. How can the bear escape? Where is the "game"? Somehow businessmen organized into "hunting clubs" forget this common sense, with the result that the lordly polar-bear is today a very seriously "endangered species." The plain sensible man understands that the sea has only so much water in it. He knows that man's population has multiplied many times in the past 200 years; that our industrial and military requirements have exploded during the same period—therefore, that man cannot continue forever to use the seas as all-purpose dump.

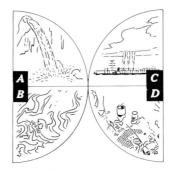

A / Industrial pollution. *Dye pouring out of a textile plant.*

B / Oil slick. *The iridescence of oil on water may look handsome, but it can destroy marine life and make our waterways unusable in the near future.*

C / Four Corners power station. *This power station, located at the desert junction of Colorado, Utah, Arizona, and New Mexico, serves Los Angeles by providing it with electricity. It also, unfortunately, pollutes the once-crystal-clear air.*

D / On a personal level. *Some residue of our twentieth-century society.*

CO₂   CO₂

FOSSIL FUEL COMBUSTION

DEAD ORGANIC MATTER AND DECOMPOSERS

MINERALS

EARTH STRATA

The Cycles of Life

Complicated as this illustration may seem, it is actually an oversimplification of the interlocking cycles on which life on earth depends. In the ocean, a large part of the organic carbon originally created by the phytoplankton ends up on the seafloor or in coral reefs in the form of calcium car-bonate—shells and skeletons of all sizes and shapes. Today the cycle is endangered. The fossil fuels are being consumed far faster than they are being replaced as a result of man's exploding industrialization of the past century, and the CO₂ content of the atmosphere has been rising in a signifi-

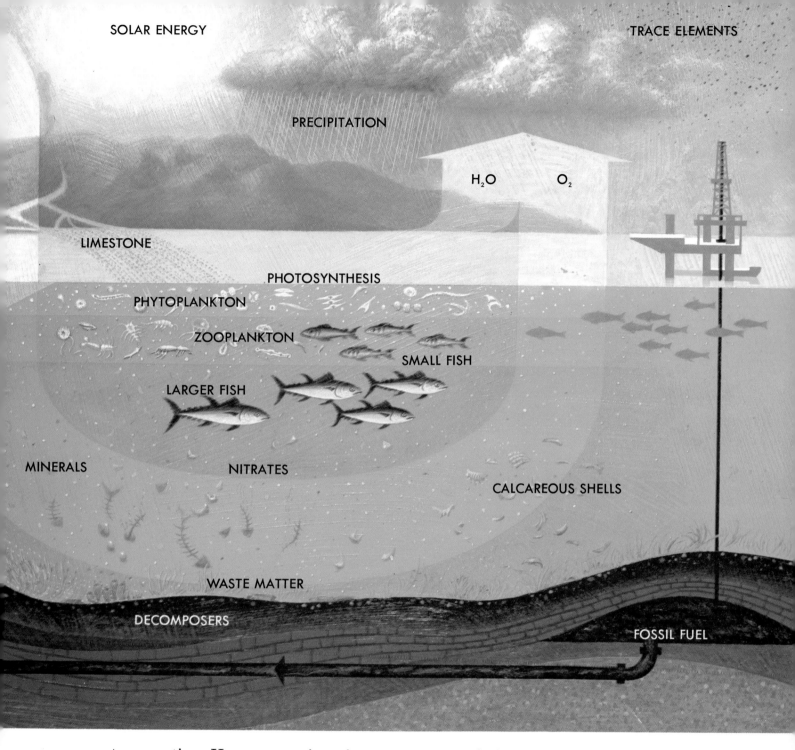

SOLAR ENERGY

TRACE ELEMENTS

PRECIPITATION

H_2O O_2

LIMESTONE

PHOTOSYNTHESIS

PHYTOPLANKTON

ZOOPLANKTON

SMALL FISH

LARGER FISH

MINERALS

NITRATES

CALCAREOUS SHELLS

WASTE MATTER

DECOMPOSERS

FOSSIL FUEL

cant proportion. However, various factors not yet clearly understood seem to be keeping the situation under control. The slowest of all cycles is that involving the many minerals which are continually contributed to the ocean as a result of the natural leaching of the land by rain and wind and of many industrial processes. Some of these mineral wastes are dissolved in seawater, more of them deposited on the sea-bed, where they join the depositing calcium carbonate in a typical bottom sediment or ooze. Here they wait for geologic changes that can take hundreds of millions of years —when at last movements of the crustal plates or volcanic upwellings squeeze them once more to the surface where they re-enter their cycle.

137

Endangered Species

One of the few animals who kills for reasons not involved in his own survival is man. We have carelessly and selfishly brought to the brink of extinction some of the most magnificent animals ever to inhabit the earth. We destroy animals for their pelts, or because they get in our way, or simply because they are there. Each animal species that disappears hastens the day when man himself will become an endangered species.

A / The polar bear. One of the largest living carnivorous land mammals. *Much larger than the females, males can weigh in considerable excess of 1000 pounds. Species exist only in and around the Arctic Ocean. There are probably 10,000 polar bears in existence, about half of which are in Canada.*

B / The Galapagos penguin. This bird cannot fly, but is an expert diver and swimmer. Its population is numbered now at less than 2400.

C / The dugong. With manatee (illustrated here), this is the only surviving representative of a once much larger order. With whales and dolphins they have "returned to the sea."

138

D / The blue whale. *This is the largest animal ever to have existed. It can be up to 120 feet long and weigh over 135 tons. One marked blue whale is known to have travelled 1900 miles in 47 days. There are probably now fewer than 1000 of these animals left. (Pictured here is a humpbacked whale.)*

E / The pelican. *The insecticide DDT unintentionally affects this bird's ability to provide the calcium needed to give its shells rigidity. The real danger to this species is seen when we realize that not all pelicans mate every year, and those which do lay only two to four eggs.*

F / The Juan Fernandez fur seal. *A distinct species known to live only on the Juan Fernandez archipelago 500 miles west of Chile. Perhaps only a few hundred survive.*

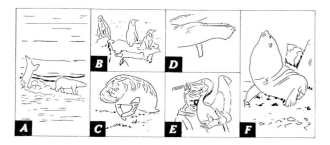

Chapter XIV. Oasis in Space

When Neil Armstrong's foot made contact with the surface of the Moon a new chapter opened in the story of the Animal World. For the first time an animal had consciously escaped from his own destiny. During all the millennia prior to Apollo 11 the propagation of species had been left to wind or water or to wings, fins, or legs. But in our own day here is man deciding on his own— and proving—that he can leave his ancestral environment—to the planet of his choice!

Our Solar System will prove to be no bargain so far as life is concerned. Scientists may have a field-day on Mars or one of Saturn's moons but none will feel homesick

"Whenever man jumps across the interstellar void, and whatever our descendants discover among the billions of stars and systems that comprise the Universe, man will never find a planet that he loves as he does earth."

for them. And we cannot even guess how many generations or centuries will pass before man, armed with instruments of propulsion now only dreams, will make the next jump—across the interstellar void in search of new suns with habitable planets. But whenever this happens, and whatever our remote descendants discover among the billions of stars and systems that comprise the Universe, man will never find a planet he can love as he can earth. Nowhere else will the breezes of a mountain-meadow at dawn so perfectly suit his lungs. Nowhere else will gravity so perfectly suit his musculature. Nowhere else will a sun's rays so perfectly suit the pigments of his skin. And

nowhere else will an ocean so perfectly resonate with the salts of his own blood. Man shares this unique habitat with the whole community of the Animal World—all of whose members are our cousins in one degree or another. Scientists tell us it is a mistake to assume that animals think and feel like humans. But the dolphin's reactions are uncannily similar to man's. The dolphin is so fast and agile that he catches his food easily. He needs companionship and affection. Unable to live alone, he has a social life in the form of strictly organized herds. Delinquency is punished by ostracism. This is the most severe punishment, as isolation means demoralization and death. He falls in love like any human adolescent, but polygamy is the rule. He (or rather she) gives birth to live young after eleven months of gestation.

In our books to follow we will explore the grand themes of this ocean that man carries with him wherever he goes. We will try to show how the instincts and drives and senses and appetites of animals and plants link up and interweave with each other, creating the colorful, complex, constantly changing tapestry of life in the sea. We must realize that man can better understand himself by understanding the sea from which he sprang. We carry more along with us on our journey through time and space than earth's ocean. Our baggage includes our zoological heritage—a heritage which, as we twentieth-century specimens know only too well, can fall dangerously out-of-tune. If we are to survive as a species, we must reconcile with nature and the sea.

Man and dolphin. *Here one of Calypso's divers hitches a ride on the dorsal fin of his "cousin," the dolphin.*

Index

ILLUSTRATIONS AND CHARTS:

Sy and Dorothea Barlowe—15, 30, 31, 33, 68, 69, 81; Walter Hortens—17, 22, 59, 64, 95, 116; Howard Koslow—33, 54, 55, 63, 92, 93, 94, 96, 97, 99, 128, 129, 133, 136, 137

PHOTO CREDITS:

Chuck Allen—83; Black Star Publications: J. Launois—131, Flip Schulke—88, Pastner—125; John Boland—52, 70, 71, 72, 73; Fred Bruemmer—84, 85; Bruce C. Coleman, Inc.: R. Thompson—39, Jane Burton—77, Jen and Des Bartlett—73, R. Mariscal—28, 101, S. Gillsater—84, 138; David Doubilet—42, 46, 79; Kingsley C. Fairbridge—19, 91; C. P. Idyll—49, 124; G. L. Kooyman——, 35; L. A. Hyperion Sewerage Plant—28; Don Lusby, Jr.—47, 85; Marineland of Florida—13; Mobil Oil Corporation: M. Anguti—124; National Audubon Society: A. W. Ambler—42, Leonard Lee Rue, III—39, Karl Kenyon—40, 41, George Lower—77, John Gerrard—131, Molly Adams—88, N. Smythe—38, 39; NASA—11, 18, 91, 115; National Marine Fishing Service—29; NUC: Steve Leatherwood—66, 67; Naval Photographic Center—11, 19, 91; Osborn Laboratories of Marine Sciences: Walter Lerchenfeld—26, 52, 77, 100, 104; Harry Pederson—104; Photo Researchers: Russ Kinne—29, 77, 83, 101, 104, Peter David—108, 109, Fred Baldwin—106, Rudolph Freund—139, George Holton—138, Joe Munroe—49; Rapho—49; F. M. Roberts, © 1966—73, 83; Dr. Roman Vishniac—53, 70, 72, 100; WHOI—108, 124; © Douglas P. Wilson—29.

The Ocean World of Jacques Cousteau

Oasis in Space